国际大学生程序设计竞赛指南
基金项目：浙江省教育厅科研项目（Y200907440）

ACM 程序设计（第 2 版）

曾棕根　编著

内 容 简 介

本书详细讲解了 ACM 国际大学生程序设计竞赛（ACM/ICPC）编程、调试方法，以及提高时间、空间性能的策略，并充分利用了 C++泛型编程的高效率、规范化的特性，全部采用 C++泛型编程。

第 1 章讲解了 ACM 程序设计入门知识；第 2 章讲解了 C++泛型编程的容器、迭代器和常用算法；第 3 章讲解了 ACM 程序设计的基本编程技巧；第 4 章讲解了 50 道原版 ACM 竞赛题的解题思路，并配有 C++泛型编程参考答案和题目的中文翻译。第 3 章和第 4 章的源代码可在 http://www.realoj.com/ 上直接下载。

本书是一本专门针对 ACM 国际大学生程序设计竞赛而编写的入门教程，适合参加 ACM/ICPC 的大学生和 C++编程爱好者学习，对 ACM/ICPC 竞赛教练也具有一定的指导作用。

图书在版编目(CIP)数据

ACM 程序设计/曾棕根编著. —2 版. —北京：北京大学出版社，2011.4
(国际大学生程序设计竞赛指南)
ISBN 978-7-301-18723-4

Ⅰ. ①A… Ⅱ. ①曾… Ⅲ. ①程序设计—教材 Ⅳ. ①TP311.1

中国版本图书馆 CIP 数据核字(2011)第 055935 号

书　　　　名：	ACM 程序设计（第 2 版）
著作责任者：	曾棕根　编著
策 划 编 辑：	胡伟晔
责 任 编 辑：	胡伟晔
标 准 书 号：	ISBN 978-7-301-18723-4/TP · 1162
出　版　者：	北京大学出版社
地　　　　址：	北京市海淀区成府路 205 号　100871
网　　　　址：	http://www.pup.cn
电　　　　话：	邮购部 62752015　发行部 62750672　编辑部 62765126　出版部 62754962
电 子 邮 箱：	zyjy@pup.cn
印　刷　者：	三河市博文印刷厂
发　行　者：	北京大学出版社
经　销　者：	新华书店
	787 毫米×1092 毫米　16 开本　18.25 印张　432 千字
	2008 年 11 月第 1 版　2011 年 4 月第 2 版　2012 年 3 月第 2 次印刷　总第 3 次印刷
定　　　　价：	34.00 元

未经许可，不得以任何方式复制或抄袭本书之部分或全部内容。

版权所有，侵权必究

举报电话：010-62752024　　　电子邮箱：fd@pup.pku.edu.cn

前　　言

ACM 国际大学生程序设计竞赛（ACM International Collegiate Programming Contest，ACM/ICPC）由 ACM（Association for Computing Machinery，美国计算机协会）主办，是世界上公认的规模最大、水平最高、影响最广的国际大学生程序设计竞赛。

由于 ACM 大学生程序设计竞赛能迅速提高学生的程序设计能力和团队协作水平，又能有效地提升学校程序设计教学水平和质量，促进校际交流与竞争，近年来，ACM 大学生程序设计竞赛在我国得到了大规模的推广，各个高校都十分重视。然而，由于该赛事是个新鲜事物，且 ACM 程序设计有其特殊的规律，很多高校难以适应。

为了尽快解决这个问题，笔者将近三年来潜心钻研 ACM 竞赛的相关资料整理出来，供大家分享。经实践证实，本书能在两个月内将程序设计的初学者培养成 ACM 竞赛高手。

本书以 C++ 泛型编程的应用为主线，一步一步带领读者登上 ACM 程序设计的殿堂。第 1 章讲解了 ACM 程序设计入门知识；第 2 章讲解了 C++ 泛型编程的容器、迭代器和常用算法；第 3 章讲解了 ACM 程序设计的基本编程技巧；第 4 章讲解了 50 道原版 ACM 竞赛题的解题思路，并配有 C++ 泛型编程参考答案和题目的中文翻译。

本书第 1 版获得了广大 ACM 竞赛师生的认可与好评，已成为大学生参加 ACM 竞赛的必备入门级读物。

本书是编者主持的浙江省教育厅科研项目（编号：Y200907440）的科研成果之一。经过近四年的深入研究，编者已在 Linux 平台上成功研发了一款高质量的 OJ 系统——RealOJ（网址 www.realoj.com）。有了上述研究背景，才有了本书的第 2 版修订。

第 2 版修订的主要地方有：

（1）采用跨平台的 Bloodshed Dev-C++ 作为编译器，书中讲述了其详细用法；

（2）采用了笔者研制的 RealOJ 平台，书上的题目在该网站里都能找到；

（3）修正了第 1 版中存在的印刷错误。

本书的出版，首先得感谢我的导师浙江师范大学数理信息学院王基一教授和丁革建教授、浙江工业大学信息学院钱能教授对笔者学术工作的不断启发与鼓励；感谢北京大学出版社的大力支持！

ACM 程序设计要求很高，由于时间仓促，笔者水平有限，疏漏之处，在所难免，恳请广大专家、读者批评指正！笔者邮箱 zjnuken@126.com，个人主页 www.realoj.com。

曾棕根
2011 年 1 月

目 录

第1章 ACM 程序设计入门 1
1.1 ACM/ICPC 简介 1
- 1.1.1 历史 1
- 1.1.2 简要规则 1
- 1.1.3 区域和全球决赛 2
- 1.1.4 历届冠军 2
- 1.1.5 源程序在线评测系统（Online Judge） 3
- 1.1.6 试题样例 4

1.2 用 Dev-C++编写控制台程序 4
- 1.2.1 例题 4
- 1.2.2 操作 4

1.3 ACM 竞赛本机调试方法 7
- 1.3.1 竞赛样题 7
- 1.3.2 本机调试步骤 8

1.4 RealOJ 源程序在线评测系统在线实验 9
- 1.4.1 竞赛样题 9
- 1.4.2 提交代码 10

第2章 C++STL 泛型编程 12
2.1 C++STL 概述 12
- 2.1.1 C++STL 的实现版本 12
- 2.1.2 C++STL 组件 12
- 2.1.3 C++STL 泛型编程示例 13
- 2.1.4 VC++ 6.0 泛型编程 14

2.2 vector 向量容器 14
- 2.2.1 创建 vector 对象 15
- 2.2.2 尾部元素扩张 15
- 2.2.3 下标方式访问 vector 元素 16
- 2.2.4 用迭代器访问 vector 元素 16
- 2.2.5 元素的插入 17
- 2.2.6 元素的删除 17
- 2.2.7 使用 reverse 反向排列算法 18
- 2.2.8 使用 sort 算法对向量元素排序 19
- 2.2.9 向量的大小 21

2.3 string 基本字符系列容器 22
- 2.3.1 创建 string 对象 22
- 2.3.2 给 string 对象赋值 22
- 2.3.3 从 string 对象尾部添加字符 23
- 2.3.4 从 string 对象尾部追加字符串 24
- 2.3.5 给 string 对象插入字符 24
- 2.3.6 访问 string 对象的元素 25
- 2.3.7 删除 string 对象的元素 25
- 2.3.8 返回 string 对象的长度 26
- 2.3.9 替换 string 对象的字符 27
- 2.3.10 搜索 string 对象的元素或子串 27
- 2.3.11 string 对象的比较 28
- 2.3.12 用 reverse 反向排序 string 对象 28
- 2.3.13 string 对象作为 vector 元素 29
- 2.3.14 string 类型的数字化处理 29
- 2.3.15 string 对象与字符数组互操作 30
- 2.3.16 string 对象与 sscanf 函数 31
- 2.3.17 string 对象与数值相互转换 32

2.4 set 集合容器 33
- 2.4.1 创建 set 集合对象 33
- 2.4.2 元素的插入与中序遍历 34
- 2.4.3 元素的反向遍历 34
- 2.4.4 元素的删除 35

		2.4.5 元素的检索 36
		2.4.6 自定义比较函数 37
2.5	multiset 多重集合容器 39	
		2.5.1 multiset 元素的插入 39
		2.5.2 multiset 元素的删除 40
		2.5.3 查找元素 41
2.6	map 映照容器 42	
		2.6.1 map 创建、元素插入和 遍历访问 42
		2.6.2 删除元素 43
		2.6.3 元素反向遍历 44
		2.6.4 元素的搜索 45
		2.6.5 自定义比较函数 45
		2.6.6 用 map 实现数字分离 47
		2.6.7 数字映照字符的 map 写法 48
2.7	multimap 多重映照容器 49	
		2.7.1 multimap 对象创建、 元素插入 49
		2.7.2 元素的删除 50
		2.7.3 元素的查找 51
2.8	deque 双端队列容器 52	
		2.8.1 创建 deque 对象 53
		2.8.2 插入元素 53
		2.8.3 前向遍历 55
		2.8.4 反向遍历 56
		2.8.5 删除元素 56
2.9	list 双向链表容器 59	
		2.9.1 创建 list 对象 60
		2.9.2 元素插入和遍历 60
		2.9.3 反向遍历 61
		2.9.4 元素删除 61
		2.9.5 元素查找 64
		2.9.6 元素排序 66
		2.9.7 剔除连续重复元素 66
2.10	bitset 位集合容器 67	
		2.10.1 创建 bitset 对象 68
		2.10.2 设置元素值 68

		2.10.3 输出元素 70
2.11	stack 堆栈容器 72	
2.12	queue 队列容器 73	
2.13	priority_queue 优先队列容器 74	
		2.13.1 优先队列的使用方法 75
		2.13.2 重载"<"操作符来 定义优先级 75
		2.13.3 重载"()"操作符来 定义优先级 77
第 3 章	ACM 程序设计基础 78	
3.1	读入一个参数 78	
		3.1.1 链接地址 78
		3.1.2 题目内容 78
		3.1.3 参考答案 78
3.2	读入两个参数 79	
		3.2.1 链接地址 79
		3.2.2 题目内容 79
		3.2.3 参考答案 79
3.3	1!到 n!的和 79	
		3.3.1 链接地址 79
		3.3.2 题目内容 80
		3.3.3 参考答案 80
3.4	等比数列 80	
		3.4.1 链接地址 80
		3.4.2 题目内容 80
		3.4.3 参考答案 81
3.5	菲波那契数 82	
		3.5.1 链接地址 82
		3.5.2 题目内容 82
		3.5.3 参考答案 82
3.6	最大公约数 83	
		3.6.1 链接地址 83
		3.6.2 题目内容 84
		3.6.3 参考答案 84
3.7	最小公倍数 85	
		3.7.1 链接地址 85
		3.7.2 题目内容 85

	3.7.3	参考答案	85
3.8	平均数		86
	3.8.1	链接地址	86
	3.8.2	题目内容	86
	3.8.3	参考答案	86
3.9	对称三位数素数		87
	3.9.1	链接地址	87
	3.9.2	题目内容	87
	3.9.3	参考答案	87
3.10	十进制转换为二进制		88
	3.10.1	链接地址	88
	3.10.2	题目内容	88
	3.10.3	参考答案	89
3.11	列出完数		89
	3.11.1	链接地址	89
	3.11.2	题目内容	89
	3.11.3	参考答案	90
3.12	12！配对		91
	3.12.1	链接地址	91
	3.12.2	题目内容	91
	3.12.3	参考答案	91
3.13	五位以内的对称素数		92
	3.13.1	链接地址	92
	3.13.2	题目内容	92
	3.13.3	参考答案	92
3.14	01串排序		93
	3.14.1	链接地址	93
	3.14.2	题目内容	93
	3.14.3	参考答案	94
3.15	排列对称串		94
	3.15.1	链接地址	94
	3.15.2	题目内容	95
	3.15.3	参考答案	95
3.16	按绩点排名		96
	3.16.1	链接地址	96
	3.16.2	题目内容	96
	3.16.3	参考答案	97
3.17	按1的个数排序		98
	3.17.1	链接地址	98
	3.17.2	题目内容	98
	3.17.3	参考答案	99

第4章 ACM程序设计实战 100

4.1	Quicksum		100
	4.1.1	链接地址	100
	4.1.2	时空限制	100
	4.1.3	题目内容	100
	4.1.4	题目来源	101
	4.1.5	解题思路	101
	4.1.6	参考答案	101
	4.1.7	汉语翻译	102
4.2	IBM Minus One		103
	4.2.1	链接地址	103
	4.2.2	时空限制	103
	4.2.3	题目内容	103
	4.2.4	题目来源	104
	4.2.5	解题思路	104
	4.2.6	参考答案	105
	4.2.7	汉语翻译	105
4.3	Binary Numbers		106
	4.3.1	链接地址	106
	4.3.2	时空限制	106
	4.3.3	题目内容	106
	4.3.4	题目来源	107
	4.3.5	解题思路	107
	4.3.6	参考答案	107
	4.3.7	汉语翻译	108
4.4	Encoding		109
	4.4.1	链接地址	109
	4.4.2	时空限制	109
	4.4.3	题目内容	109
	4.4.4	题目来源	109
	4.4.5	解题思路	109
	4.4.6	参考答案	110
	4.4.7	汉语翻译	111

- 4.5 Look and Say 111
  - 4.5.1 链接地址 111
  - 4.5.2 时空限制 111
  - 4.5.3 题目内容 111
  - 4.5.4 题目来源 112
  - 4.5.5 解题思路 112
  - 4.5.6 参考答案 112
  - 4.5.7 汉语翻译 113
- 4.6 Abbreviation 114
  - 4.6.1 链接地址 114
  - 4.6.2 时空限制 114
  - 4.6.3 题目内容 114
  - 4.6.4 题目来源 115
  - 4.6.5 解题思路 115
  - 4.6.6 参考答案 116
  - 4.6.7 汉语翻译 116
- 4.7 The Seven Percent Solution 118
  - 4.7.1 链接地址 118
  - 4.7.2 时空限制 118
  - 4.7.3 题目内容 118
  - 4.7.4 题目来源 119
  - 4.7.5 解题思路 119
  - 4.7.6 参考答案 119
  - 4.7.7 汉语翻译 120
- 4.8 Digital Roots 121
  - 4.8.1 链接地址 121
  - 4.8.2 时空限制 121
  - 4.8.3 题目内容 121
  - 4.8.4 题目来源 122
  - 4.8.5 解题思路 122
  - 4.8.6 参考答案 122
  - 4.8.7 汉语翻译 123
- 4.9 Box of Bricks 124
  - 4.9.1 链接地址 124
  - 4.9.2 时空限制 124
  - 4.9.3 题目内容 124
  - 4.9.4 题目来源 125
  - 4.9.5 解题思路 125
  - 4.9.6 参考答案 125
  - 4.9.7 汉语翻译 126
- 4.10 Geometry Made Simple 127
  - 4.10.1 链接地址 127
  - 4.10.2 时空限制 127
  - 4.10.3 题目内容 127
  - 4.10.4 题目来源 128
  - 4.10.5 解题思路 128
  - 4.10.6 参考答案 129
  - 4.10.7 汉语翻译 130
- 4.11 Reverse Text 131
  - 4.11.1 链接地址 131
  - 4.11.2 时空限制 131
  - 4.11.3 题目内容 131
  - 4.11.4 题目来源 132
  - 4.11.5 解题思路 132
  - 4.11.6 参考答案 132
  - 4.11.7 汉语翻译 133
- 4.12 Word Reversal 133
  - 4.12.1 链接地址 133
  - 4.12.2 时空限制 133
  - 4.12.3 题目内容 133
  - 4.12.4 题目来源 134
  - 4.12.5 解题思路 134
  - 4.12.6 参考答案 134
  - 4.12.7 汉语翻译 135
- 4.13 A Simple Question of Chemistry 136
  - 4.13.1 链接地址 136
  - 4.13.2 时空限制 136
  - 4.13.3 题目内容 136
  - 4.13.4 题目来源 137
  - 4.13.5 解题思路 137
  - 4.13.6 参考答案 137
  - 4.13.7 汉语翻译 138
- 4.14 Adding Reversed Numbers 139
  - 4.14.1 链接地址 139
  - 4.14.2 时空限制 139
  - 4.14.3 题目内容 139

- 4.14.4 题目来源 140
- 4.14.5 解题思路 140
- 4.14.6 参考答案 140
- 4.14.7 汉语翻译 142
- 4.15 Image Transformation 143
  - 4.15.1 链接地址 143
  - 4.15.2 时空限制 143
  - 4.15.3 题目内容 143
  - 4.15.4 题目来源 144
  - 4.15.5 解题思路 144
  - 4.15.6 参考答案 144
  - 4.15.7 汉语翻译 145
- 4.16 Beautiful Meadow 146
  - 4.16.1 链接地址 146
  - 4.16.2 时空限制 146
  - 4.16.3 题目内容 147
  - 4.16.4 题目来源 148
  - 4.16.5 解题思路 148
  - 4.16.6 参考答案 148
  - 4.16.7 汉语翻译 150
- 4.17 DNA Sorting 151
  - 4.17.1 链接地址 151
  - 4.17.2 时空限制 151
  - 4.17.3 题目内容 151
  - 4.17.4 题目来源 152
  - 4.17.5 解题思路 152
  - 4.17.6 参考答案 152
  - 4.17.7 汉语翻译 154
- 4.18 Daffodil Number 155
  - 4.18.1 链接地址 155
  - 4.18.2 时空限制 155
  - 4.18.3 题目内容 155
  - 4.18.4 题目来源 155
  - 4.18.5 解题指导 155
  - 4.18.6 参考答案 156
  - 4.18.7 汉语翻译 156
- 4.19 Error Correction 157
  - 4.19.1 链接地址 157
- 4.19.2 时空限制 157
- 4.19.3 题目内容 157
- 4.19.4 题目来源 158
- 4.19.5 解题思路 158
- 4.19.6 参考答案 158
- 4.19.7 汉语翻译 160
- 4.20 Martian Addition 161
  - 4.20.1 链接地址 161
  - 4.20.2 时空限制 161
  - 4.20.3 题目内容 161
  - 4.20.4 题目来源 162
  - 4.20.5 解题思路 162
  - 4.20.6 参考答案 162
  - 4.20.7 汉语翻译 164
- 4.21 FatMouse' Trade 165
  - 4.21.1 链接地址 165
  - 4.21.2 时空限制 165
  - 4.21.3 题目内容 165
  - 4.21.4 题目来源 166
  - 4.21.5 解题指导 166
  - 4.21.6 参考答案 166
  - 4.21.7 汉语翻译 168
- 4.22 List the Books 168
  - 4.22.1 链接地址 168
  - 4.22.2 时空限制 169
  - 4.22.3 题目内容 169
  - 4.22.4 题目来源 170
  - 4.22.5 解题指导 170
  - 4.22.6 参考答案 170
  - 4.22.7 汉语翻译 172
- 4.23 Head-to-Head Match 173
  - 4.23.1 链接地址 173
  - 4.23.2 时空限制 173
  - 4.23.3 题目内容 173
  - 4.23.4 题目来源 174
  - 4.23.5 解题指导 174
  - 4.23.6 参考答案 174
  - 4.23.7 汉语翻译 175

4.24 Windows Message Queue 175	4.28.6 参考答案 189
4.24.1 链接地址 175	4.28.7 汉语翻译 191
4.24.2 时空限制 175	4.29 Semi-Prime 192
4.24.3 题目内容 176	4.29.1 链接地址 192
4.24.4 题目来源 176	4.29.2 时空限制 192
4.24.5 解题指导 176	4.29.3 题目内容 192
4.24.6 参考答案 177	4.29.4 题目来源 193
4.24.7 汉语翻译 178	4.29.5 解题思路 193
4.25 Language of FatMouse 178	4.29.6 参考答案 193
4.25.1 链接地址 178	4.29.7 汉语翻译 194
4.25.2 时空限制 179	4.30 Beautiful Number 195
4.25.3 题目内容 179	4.30.1 链接地址 195
4.25.4 题目来源 179	4.30.2 时空限制 195
4.25.5 解题思路 179	4.30.3 题目内容 195
4.25.6 参考答案 180	4.30.4 题目来源 196
4.25.7 汉语翻译 181	4.30.5 解题思路 196
4.26 Palindromes 181	4.30.6 参考答案 196
4.26.1 链接地址 181	4.30.7 汉语翻译 197
4.26.2 时空限制 182	4.31 Phone List 198
4.26.3 题目内容 182	4.31.1 链接地址 198
4.26.4 题目来源 182	4.31.2 时空限制 198
4.26.5 解题思路 182	4.31.3 题目内容 198
4.26.6 参考答案 183	4.31.4 题目来源 199
4.26.7 汉语翻译 185	4.31.5 解题思路 199
4.27 Root of the Problem 185	4.31.6 参考答案 199
4.27.1 链接地址 185	4.31.7 汉语翻译 202
4.27.2 时空限制 185	4.32 Calendar 203
4.27.3 题目内容 185	4.32.1 链接地址 203
4.27.4 题目来源 186	4.32.2 时空限制 203
4.27.5 解题思路 186	4.32.3 题目内容 203
4.27.6 参考答案 186	4.32.4 题目来源 204
4.27.7 汉语翻译 187	4.32.5 解题思路 204
4.28 Magic Square 187	4.32.6 参考答案 204
4.28.1 链接地址 187	4.32.7 汉语翻译 206
4.28.2 时空限制 188	4.33 No Brainer 207
4.28.3 题目内容 188	4.33.1 链接地址 207
4.28.4 题目来源 189	4.33.2 时空限制 207
4.28.5 解题思路 189	4.33.3 题目内容 207

目 录

4.33.4 题目来源 208	4.38.2 时空限制 223
4.33.5 解题思路 208	4.38.3 题目内容 223
4.33.6 参考答案 208	4.38.4 题目来源 224
4.33.7 汉语翻译 209	4.38.5 解题思路 224
4.34 Quick Change 209	4.38.6 参考答案 224
4.34.1 链接地址 209	4.38.7 汉语翻译 225
4.34.2 时空限制 209	4.39 Champion of the Swordsmanship ... 226
4.34.3 题目内容 210	4.39.1 链接地址 226
4.34.4 题目来源 210	4.39.2 时空限制 226
4.34.5 解题思路 210	4.39.3 题目内容 226
4.34.6 参考答案 211	4.39.4 题目来源 227
4.34.7 汉语翻译 211	4.39.5 解题思路 227
4.35 Total Amount 212	4.39.6 参考答案 227
4.35.1 链接地址 212	4.39.7 汉语翻译 228
4.35.2 时空限制 212	4.40 Doubles 228
4.35.3 题目内容 212	4.40.1 链接地址 228
4.35.4 题目来源 213	4.40.2 时空限制 229
4.35.5 解题思路 213	4.40.3 题目内容 229
4.35.6 参考答案 213	4.40.4 题目来源 229
4.35.7 汉语翻译 216	4.40.5 解题思路 229
4.36 Electrical Outlets 216	4.40.6 参考答案 230
4.36.1 链接地址 216	4.40.7 汉语翻译 230
4.36.2 时空限制 217	4.41 File Searching 231
4.36.3 题目内容 217	4.41.1 链接地址 231
4.36.4 题目来源 218	4.41.2 时空限制 231
4.36.5 解题思路 218	4.41.3 题目内容 231
4.36.6 参考答案 218	4.41.4 题目来源 233
4.36.7 汉语翻译 219	4.41.5 解题思路 233
4.37 Speed Limit 220	4.41.6 参考答案 234
4.37.1 链接地址 220	4.41.7 汉语翻译 236
4.37.2 时空限制 220	4.42 Old Bill 237
4.37.3 题目内容 220	4.42.1 链接地址 237
4.37.4 题目来源 221	4.42.2 时空限制 237
4.37.5 解题思路 221	4.42.3 题目内容 237
4.37.6 参考答案 221	4.42.4 题目来源 238
4.37.7 汉语翻译 222	4.42.5 解题思路 238
4.38 Beat the Spread! 223	4.42.6 参考答案 238
4.38.1 链接地址 223	4.42.7 汉语翻译 240

4.43	Divisor Summation 241		4.47.3	题目内容 255
	4.43.1 链接地址 241		4.47.4	题目来源 256
	4.43.2 时空限制 241		4.47.5	解题思路 256
	4.43.3 题目内容 241		4.47.6	参考答案 257
	4.43.4 题目来源 241		4.47.7	汉语翻译 258
	4.43.5 解题思路 242	4.48	Excuses, Excuses! 259
	4.43.6 参考答案 242		4.48.1 链接地址 259
	4.43.7 汉语翻译 243		4.48.2 时空限制 259
4.44	Easier Done Than Said? 243		4.48.3 题目内容 259
	4.44.1 链接地址 243		4.48.4 题目来源 261
	4.44.2 时空限制 243		4.48.5 解题思路 261
	4.44.3 题目内容 243		4.48.6 参考答案 261
	4.44.4 题目来源 244		4.48.7 汉语翻译 263
	4.44.5 解题思路 245	4.49	Lowest Bit 265
	4.44.6 参考答案 245		4.49.1 链接地址 265
	4.44.7 汉语翻译 247		4.49.2 时空限制 265
4.45	Let the Balloon Rise 248		4.49.3 题目内容 265
	4.45.1 链接地址 248		4.49.4 题目来源 265
	4.45.2 时空限制 248		4.49.5 解题思路 266
	4.45.3 题目内容 248		4.49.6 参考答案 266
	4.45.4 题目来源 249		4.49.7 汉语翻译 266
	4.45.5 解题思路 249	4.50	Longest Ordered Subsequence 267
	4.45.6 参考答案 249		4.50.1 链接地址 267
	4.45.7 汉语翻译 250		4.50.2 时空限制 267
4.46	The Hardest Problem Ever 251		4.50.3 题目内容 267
	4.46.1 链接地址 251		4.50.4 题目来源 268
	4.46.2 时空限制 251		4.50.5 解题思路 268
	4.46.3 题目内容 251		4.50.6 参考答案 269
	4.46.4 题目来源 252		4.50.7 汉语翻译 271
	4.46.5 解题思路 252	附录 1	用 VC++编写控制台程序的方法 272
	4.46.6 参考答案 252		
	4.46.7 汉语翻译 254	附录 2	本书试题第三方 ACM 网站链接 277
s4.47	Fibonacci Again 255		
	4.47.1 链接地址 255	参考文献	 280
	4.47.2 时空限制 255		

第 1 章　ACM 程序设计入门

1.1　ACM/ICPC 简介

ACM 国际大学生程序设计大赛（ACM International Collegiate Programming Contest，ACM/ICPC）是由世界计算机权威组织——ACM（Association for Computing Machinery，美国计算机协会）主办，是世界上公认的规模最大、水平最高的国际大学生程序设计竞赛，素来被冠以"程序设计的奥林匹克"的尊称。

1.1.1　历史

竞赛的历史可以上溯到 1970 年，当时在美国得克萨斯 A&M 大学举办了首届比赛。当时的主办方是 the Alpha Chapter of the UPE Computer Science Honor Society。作为一种全新的发现和培养计算机科学顶尖学生的方式，竞赛很快得到美国和加拿大各大学的积极响应。1977 年，在 ACM 计算机科学会议期间举办了首次总决赛，并演变成为目前的一年一届的多国参与的国际性比赛。

最初几届比赛的参赛队伍主要来自美国和加拿大，后来逐渐发展成为一项世界范围内的竞赛。特别是自 1997 年 IBM 开始赞助赛事之后，赛事规模增长迅速。1997 年，总共有来自 560 所大学的 840 支队伍参加比赛。而到了 2004 年，这一数字迅速增加到 840 所大学的 4109 支队伍并以每年 10%~20%的速度在增长。

1980 年，ACM 将竞赛的总部设在位于美国得克萨斯州的贝勒大学。

在赛事的早期，冠军多为美国和加拿大的大学获得。而进入 20 世纪 90 年代后期以来，俄罗斯和其他一些东欧国家的大学连夺数次冠军。来自中国大陆的上海交通大学代表队则在 2002 年美国夏威夷第 26 届和 2005 年上海举行的第 29 届全球总决赛上两次夺得冠军。这也是目前为止亚洲大学在该竞赛上取得的最好成绩。赛事的竞争格局已经由最初的北美大学的一枝独秀演变成目前的亚欧对抗的局面。

1.1.2　简要规则

ACM/ICPC 以团队的形式代表各学校参赛，每队由 3 名队员组成。每位队员必须是入校 5 年内的在校大学生，最多可以参加 2 次全球总决赛和 4 次区域选拔赛。比赛期间，每队使用 1 台电脑，在 5 个小时内使用 C、C++、Pascal 或 Java 中的一种语言编写程序解决 8 或 10 个问题（区域选拔赛通常是 8 题，全球总决赛是 10 题）。程序完成之后提交给在线评测（Online Judge，OJ）系统去运行，运行的结果会判定为正确或错误并及时通知参赛队。每队在正确完成一题后，组织者将在其位置上升起一只代表该题颜色的气球。

最后的获胜者为正确解答题目最多且总用时最少的队伍。每道试题用时将从竞赛开始到试题解答被判定为正确为止，其间每一次提交运行结果被判错误的话将被加罚 20 分钟时间，未正确解答的试题不记时。例如：A、B 两队都正确完成两道题目，其中 A 队提交这两题的时间分别是比赛开始后 1h 和 2h45min，B 队为 1h20min 和 2h，但 B 队有一题提交了两次（其中一次是错误的提交）。这样 A 队的总用时为 3h45min，而 B 队为 3h40min，所以 B 队以总用时少而获胜。

1.1.3 区域和全球决赛

与其他计算机程序竞赛（例如国际信息学奥林匹克，IOI）相比，ACM/ICPC 的特点在于其题量大，每队需要 5 小时内完成 8 道题目，甚至更多。另外一支队伍 3 名队员却只有 1 台电脑，使得时间显得更为紧张。因此除了扎实的专业水平外，良好的团队协作和心理素质同样是获胜的关键。

赛事由各大洲区域预赛和全球总决赛两个阶段组成。各预赛区第一名自动获得参加全球总决赛的资格。决赛安排在每年的 3～4 月举行，而区域预赛一般安排在上一年的 9～12 月举行。一个大学可以有多支队伍参加区域预赛，但只能有一支队伍参加全球总决赛。

全球总决赛第一名将获得奖杯一座。另外，成绩靠前的参赛队伍也将获得金、银和铜牌。而解题数在中等以下的队伍会得到确认但不会进行排名。

1.1.4 历届冠军

下表列出了自 1977 年以来，截至 2005 年历年全球总决赛的冠军。

ACM 竞赛历年全球冠军

年份	总决赛地点	冠军大学	国家
2005	中国上海	上海交通大学	中国
2004	捷克布拉格	圣彼得堡光学与精密仪器学院	俄罗斯
2003	美国洛杉矶	华沙大学	波兰
2002	美国夏威夷	上海交通大学	中国
2001	加拿大温哥华	国立圣彼得堡大学	俄罗斯
2000	美国奥兰多	国立圣彼得堡大学	俄罗斯
1999	荷兰爱因霍温	滑铁卢大学	加拿大
1998	美国亚特兰大	查尔斯大学	捷克
1997	美国圣何塞	哈维玛德大学	美国
1996	美国费城	加州大学伯克利分校	美国
1995	美国纳什维尔	Albert-Ludwigs-Universitat Freiburg（弗莱堡大学）	德国

（续表）

年份	总决赛地点	冠军大学	国家
1994	美国菲尼克斯	滑铁卢大学	加拿大
1993	美国印第安纳波利斯	哈佛大学	美国
1992	美国堪萨斯城	墨尔本大学	澳大利亚
1991	美国圣安东尼奥	斯坦福大学	美国
1990	美国华盛顿	奥塔哥大学	新西兰
1989	美国路易斯维尔	加州大学洛杉矶分校	美国
1988	美国亚特兰大	加州理工学院	美国
1987	美国圣路易斯	斯坦福大学	美国
1986	美国辛辛那提	加州理工学院	美国
1985	美国新奥尔良	斯坦福大学	美国
1984	美国费城	约翰霍普金斯大学	美国
1983	美国墨尔本	内布拉斯加大学	美国
1982	美国印第安纳波利斯	贝勒大学	美国
1981	美国圣路易斯	密苏里大学罗拉分校	美国
1980	美国堪萨斯城	华盛顿大学圣路易斯分校	美国
1979	美国代顿	华盛顿大学圣路易斯分校	美国
1978	美国底特律	麻省理工学院	美国
1977	美国亚特兰大	密歇根州立大学	美国

1.1.5 源程序在线评测系统（Online Judge）

源程序在线评测（Online Judge，OJ）系统上有大量的试题，只需在 OJ 系统上免费注册一个账号即可做题。

竞赛试题涵盖的范围很广，大致划分如下：Direct（简单题），Computational Geometry（计算几何），Number Theory（数论），Combinatorics（组合数学），Search Techniques（搜索技术），Dynamic Programming（动态规划），Graph Theory（图论），Other（其他）。

比较著名的 OJ 系统有：

RealOJ：http://www.realoj.com，中文题目较多，题目增加速度很快，特别适合初学者。

浙江大学 ZOJ：http://acm.zju.edu.cn/，是国内最早的 OJ，题目数量众多且质量较高。

俄罗斯乌拉尔大学的 ACM 网站：http://acm.timus.ru，是一个老牌的 OJ，题目不多，但比较经典。

ICPC 官方网站：http://cm.baylor.edu/welcome.icpc，上面会公布每年世界总决赛的排名及试题。

1.1.6 试题样例

题目名称：A + B Problem

链接地址：http://www.realoj.com /网上第 1 题

Time Limit: 1000 ms Resident Memory Limit: 1024 KB Output Limit: 1024 B

Calculate a + b

Input

The input will consist of a series of pairs of integers a and b, separated by a space, one pair of integers per line.

Output

For each pair of input integers a and b you should output the sum of a and b in one line, and with one line of output for each line in input.

Sample Input

```
1 5
```

Sample Output

```
6
```

Hint

```
Use + operator
```

1.2 用 Dev-C++编写控制台程序

在 ACM 竞赛中，一般使用 C++语言来编制程序。C++编译器有很多，笔者推荐 Bloodshed Dev-C++，它是一个跨平台的编译器，RealOJ 判题服务器上就是使用该编译器，可直接登录到 RealOJ 系统中去下载（www.realoj.com）。

本书的程序都是采用 Bloodshed Dev-C++来编写的。下面讲讲 Bloodshed Dev-C++的控制台程序的编写方法。

1.2.1 例题

编制一个 C++程序，输入 a 和 b 两个整数，输出这两个整数的和。

1.2.2 操作

（1）运行 Bloodshed Dev-C++，单击工具栏上第 1 排第 3 个按钮，就建立了一个新的源程序，如图 1-1 所示。

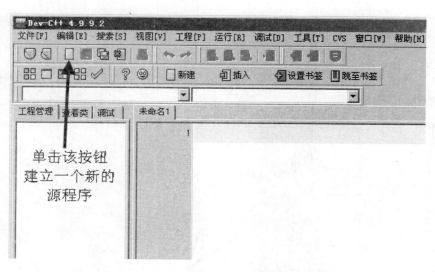

图 1-1　新建一个源程序

（2）输入 C++源代码，如图 1-2 所示。

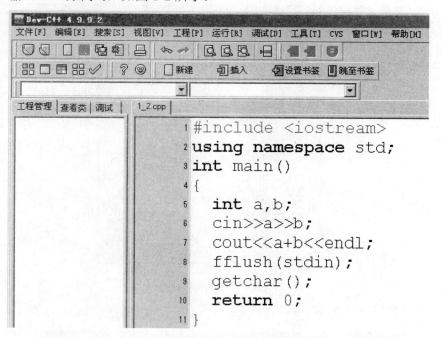

图 1-2　输入 C++源程序

（3）单击工具栏第 2 排第 1 个按钮编译工程，再在"保存文件"对话框中输入工程名称"1_2"，如图 1-3 所示。

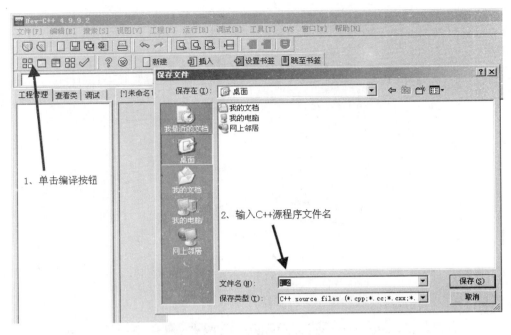

图 1-3　输入工程名称 "1_2"

（4）单击"保存"按钮，再弹出编译成功的消息框，如图 1-4 所示。

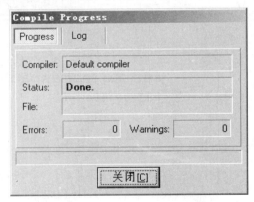

图 1-4　编译成功的消息框

（5）单击工具栏第 2 排第 3 个按钮 ▦ 后，程序自动运行了，如图 1-5 所示。

图 1-5　程序自动运行

小提示

C++类都在 std 命名空间中，所以，如果是编写 C++程序，那么，都需要使用"using namespace std;"语句来声明程序中的 C++类是在 std 命名空间中，否则，程序会出现编译错误。

标准输入流对象 cin 和标准输出流对象 cout 在头文件 iostream 中定义了"extern _CRTIMP istream cin;"和"extern_CRTIMP ostream cout;"，所以需要头文件包含声明"#include <iostream>"。cin 默认的对象是键盘设备，cout 默认的对象是屏幕设备。

另外，包含 C++文件的方法都是采用"#include <iostream>"的形式。C++类文件名都不带".h"，而带".h"的头文件名称都是 C 语言的。

1.3 ACM 竞赛本机调试方法

在 ACM 竞赛中，在将编制好的程序提交到源程序在线评测系统（Online Judge）以前，必须在本机上调试通过。

在本机调试的方法比较讲究，一般是从一个文本文件，如"aaa.txt"中读入数据，再输出数据到屏幕上即可。本书以一例说明本机调试的步骤。

1.3.1 竞赛样题

题目名称：A + B Problem

链接地址：http://www.realoj.com/网上第 1 题

Time Limit: 1000 ms   Resident Memory Limit: 1024 KB   Output Limit: 1024 B

Calculate a + b

Input

The input will consist of a series of pairs of integers a and b,separated by a space, one pair of integers per line.

Output

For each pair of input integers a and b you should output the sum of a and b in one line,and with one line of output for each line in input.

Sample Input

1 5

Sample Output

6

Hint

```
Use + operator
```

1.3.2 本机调试步骤

（1）在桌面上建立文件夹 1_3，并运行 Dev-CPP 新建一个源程序 1_3.cpp，保存在桌面上建立的文件夹 1_3 中。

（2）在桌面的 1_3 文件夹下新建一个文本文件 aaa.txt，并在该文件中输入一些测试数据（数据间用空隔或回车隔开都可以）：1 5 3 9 2 3 5 6 7 9，如图 1-6 所示。

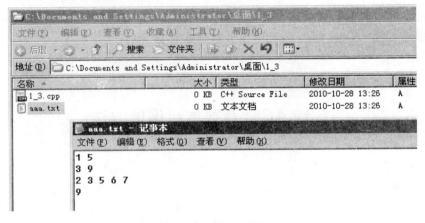

图 1-6　aaa.txt 文件内容

（3）关闭记事本，并保存对 aaa.txt 文件的修改，再编写代码，如图 1-7 所示。

图 1-7　编写程序代码

ACM 程序设计入门　第 1 章

> **小提示**
>
> ifstream 文件流类在 fstream 中定义了，所以需要头文件包含"#include <fstream>"；另外，输入对象 cin 被设置为 aaa.txt 文件了，所以，程序运行时，会自动从 aaa.txt 文件中读入数据。
>
> 注意，cin 在读入数据时，会忽略空格和回车符。

（4）按 Ctrl + F9 组合键编译，再按 F9 键运行程序，在控制台中，立即显示了运算结果，如图 1-8 所示。

图 1-8　控制台

（5）按任意键，退出控制台。本机调试完成后，就直接将代码提交到源程序在线评测系统即可，在线评测系统的使用方法在下一节会作详细的介绍。

>
>
> 使用文本文件读入数据的调试方式，使得数据自动被程序读取，另外，测试数据也可以随时修改，这样的测试方法显得很方便。

1.4　RealOJ 源程序在线评测系统在线实验

本机调试好后，就要把代码提交到源程序在线评测系统中去了。
本节详细介绍在 RealOJ 源程序在线评测系统上在线提交代码。

1.4.1　竞赛样题

题目名称：A + B Problem
链接地址：http://www.realoj.com/网上第 1 题
Time Limit: 1000 ms　Resident Memory Limit: 1024 KB　Output Limit: 1024 B
Calculate a + b

Input

The input will consist of a series of pairs of integers a and b,separated by a space, one pair of integers per line.

Output

For each pair of input integers a and b you should output the sum of a and b in one line, and with one line of output for each line in input.

Sample Input

```
1 5
```

Sample Output

```
6
```

Hint

```
Use + operator
```

1.4.2 提交代码

（1）在 www.realoj.com 第 1 题的下方，选择源代码的语言为 C++，再打开 1.3 节的工程文件，从 Dev-CPP 中把代码复制到编辑窗中，如图 1-9 所示。

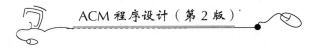

图 1-9 复制源代码

注意：一定要把"ifstream cin("aaa.txt");"这句注释掉。

（2）单击提交代码按钮，稍微等一到两秒，会弹出状态页。状态页上显示了刚才提交的是否通过（Accepted），如果 Judge Status 是 Accepted，则表明这道题做对了。如图 1-10 所示。

ACM 程序设计入门 第1章

总记录数:6509条,总页数:326页,每页显示:20条记录,当前页:1   下一页   最后页

运行ID	题目ID	源代码	用户ID	账号	编译器	判题结果	时间	可写内存	输出	提交时间
1288244636	1	查看	查看	3	zzgtest	C++	Accepted	0 ms	92 KB	9 byte	2010-10-28 13:44:00
1288239201	28	查看	N/A	233	计算机3101迷失特陈	C	Accepted	0 ms	64 KB	2 byte	2010-10-28 12:13:23
1288237463	27	查看	N/A	233	计算机3101迷失特陈	C	Accepted	0 ms	68 KB	8 byte	2010-10-28 11:44:25
1288236694	24	查看	N/A	64	计算机3101陈义勇	C	Accepted	0 ms	64 KB	19 byte	2010-10-28 11:31:37
1288236437	25	查看	N/A	64	计算机3101陈义勇	C	Accepted	0 ms	64 KB	14 byte	2010-10-28 11:27:19
1288236329	25	查看	N/A	195	110	C	Accepted	10 ms	64 KB	14 byte	2010-10-28 11:25:31
1288236097	24	查看	N/A	195	110	C	Wrong Answer	0 ms	64 KB	19 byte	2010-10-28 11:21:39
1288235729	24	查看	N/A	195	110	C	Wrong Answer	0 ms	68 KB	19 byte	2010-10-28 11:15:32
1288235623	24	查看	N/A	195	110	C	Wrong Answer	0 ms	64 KB	19 byte	2010-10-28 11:13:46
1288235250	25	查看	N/A	233	计算机3101迷失特陈	C	Accepted	10 ms	68 KB	14 byte	2010-10-28 11:07:33
1288235238	24	查看	N/A	195	110	C	Accepted	0 ms	68 KB	19 byte	2010-10-28 11:07:20
1288234754	24	查看	N/A	195	110	C	Wrong Answer	0 ms	68 KB	19 byte	2010-10-28 10:59:16
1288234726	25	查看	N/A	233	计算机3101迷失特陈	C	Wrong Answer	0 ms	64 KB	14 byte	2010-10-28 10:58:48
1288234565	24	查看	N/A	195	110	C	Wrong Answer	0 ms	64 KB	19 byte	2010-10-28 10:56:08
1288234109	25	查看	N/A	233	计算机3101迷失特陈	C	Wrong Answer	0 ms	64 KB	14 byte	2010-10-28 10:48:31
1288233536	24	查看	N/A	38	计算机3101尤森丰	C	Wrong Answer	0 ms	60 KB	19 byte	2010-10-28 10:38:58
1288232592	24	查看	N/A	38	计算机3101尤森丰	C	Presentation Error	0 ms	64 KB	20 byte	2010-10-28 10:23:15
1288232432	10	查看	N/A	148	计算机3105朱荣杰	C	Accepted	0 ms	64 KB	44 byte	2010-10-28 10:20:34
1288232236	10	查看	N/A	148	计算机3105朱荣杰	C	Wrong Answer	0 ms	60 KB	42 byte	2010-10-28 10:17:18
1288232191	24	查看	N/A	38	计算机3101尤森丰	C	Presentation Error	0 ms	68 KB	20 byte	2010-10-28 10:16:33

图 1-10 状态页面

小提示

状态页面显示了运行 ID、题目 ID、源代码、用户 ID、账号、编译器、判题结果、时间、可写内存、输出、提交时间等几个信息。判题结果一般有以下几种情况。

Accepted：恭喜，通过。

Presentation Error：输出格式错误。

Wrong Answer：答案错误。

Runtime Error：程序运行时发生错误，多为数组访问越界。

Time Limit Exceeded：超时错误，程序运行时间超过限制的时间。

Memory Limit Exceeded：超内存错误，程序运行使用的内存超过限制的内存用量。

Compile Error：编译错误，源代码中有语法错误。

Empty Anstwer：答案为空。

Output Limit Exceeded：超输出错误，程序输出数据量超限。

第 2 章　C++STL 泛型编程

2.1　C++STL 概述

在 ACM 竞赛中，需要用到数组、字符串、队列、堆栈、链表、平衡二叉检索树等数据结构和排序、搜索等算法，以提高程序的时间、空间运行效率。这些数据结构，如果都需要手工来编写，那是相当麻烦的事情。

幸运的是，ANSI C++中包含了一个 C++ STL（Standard Template Library），即 C++标准模板库，又称 C++泛型库，它在 std 命名空间中定义了常用的数据结构和算法，使用起来十分方便。

2.1.1　C++STL 的实现版本

C++STL 的实现版本主要有 HP STL、SGI STL、STLport、P.J.Plauger STL 和 Rouge Wave STL 等五种。

HP STL 是 Alexandar Stepanov（1950 年生于莫斯科，被称为 STL 之父）在惠普 Palo Alto 实验室工作时，与 Meng Lee 合作完成的，它是 C++STL 的第一个实现版本，而且是开放源码，其他版本的 C++STL，一般是以 HP STL 为蓝本来实现的。

SGI STL 是由 Silicon Graphics Computer Systems 公司参照 HP STL 实现的，主要设计者仍然是 Alexandar Stepanov，被 Linux 的 C++编译器 GCC 所采用。SGI STL 是开源软件，可以在 http://www.sgi.com/网站上下载。

STLport 是俄国人 Boris Fomitchev 建立的一个 free 项目，它是开放源码的，主要目的是使 SGI STL 的基本代码都适用于 VC++和 C++Builder 等多种编译器，源码可以在 http://www.stlport.com/ 网站上下载。

P.J.Plauger STL 是由 P.J.Plauger 参照 HP STL 实现的，被 Visual C++编译器所采用，它不是开源的。头文件在 C:\Program Files\Microsoft Visual Studio\VC98\Include 文件夹下，文件名不带".h"。

Rouge Wave STL 是 Rouge Wave 公司参照 HP STL 实现的，用于 Borland C++编译器中，当然，它也不是开源的。

2.1.2　C++STL 组件

STL 提供三种类型的组件：容器、迭代器和算法，它们都支持泛型程序设计标准。

容器主要有两类：顺序容器和关联容器。顺序容器（vector、list、deque 和 string 等）是一系列元素的有序集合。关联容器（set、multiset、map 和 multimap）包含查找元素的键值。

迭代器的作用是遍历容器。

STL 算法库包含四类算法：排序算法、不可变序算法、变序性算法和数值算法。

2.1.3 C++STL 泛型编程示例

用 vector 向量容器装入 10 个整数，然后，使用迭代器 iterator 和 accumulate 算法统计出这 10 个元素的和。

采用 VC++6.0 编制的泛型程序如下：

```
#include "stdafx.h"
//cin 和 cout 需要
#include <iostream>
//向量需要
#include <vector>
//accumulate 算法需要
#include <numeric>
using namespace std;
int main(int argc, char * argv[])
{
    //定义向量 v
    vector<int> v;
    int i;
    //赋值
    for(i=0;i<10;i++)
    {
        //尾部元素扩张方式赋值
        v.push_back(i);
    }
    //使用 iterator 迭代器顺序遍历所有元素
    for(vector<int>::iterator it=v.begin();it!=v.end();it++)
    {
        //输出迭代器当前位置上的元素值
        cout<<*it<<" ";
    }
    cout<<endl;//回车换行
    //统计并输出向量所有元素的和
    cout<<accumulate(v.begin(),v.end(),0)<<endl;
    return 0;
}
```

程序运行后的结果如图 2-1 所示。

图 2-1 程序运行结果

2.1.4　VC++ 6.0 泛型编程

在 Microsoft Visual C++ 6.0 中，C++STL 泛型头文件在 C:\Program Files\Microsoft Visual Studio\VC98\Include 文件夹下，文件名中不带".h"的文件都是，如图 2-2 所示。

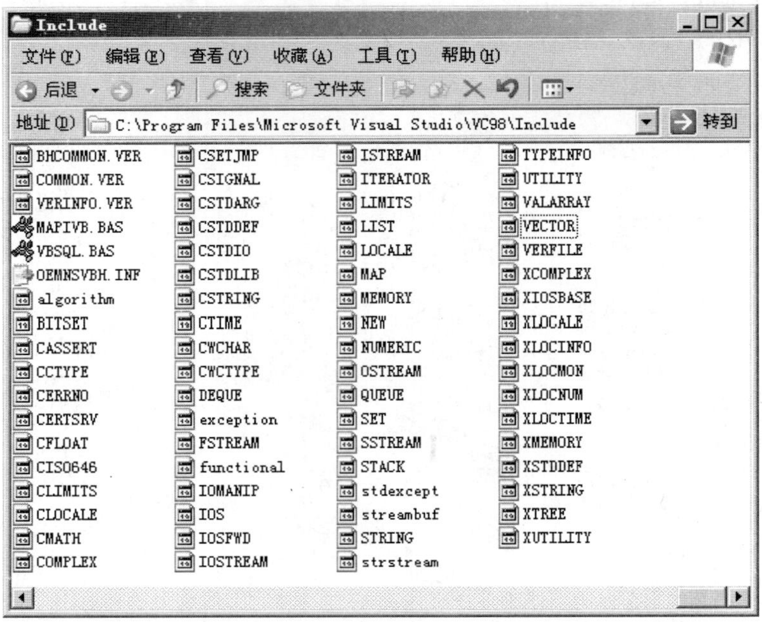

图 2-2　VC++ 6.0 中的 C++STL 泛型头文件

如果要使用 vector 向量容器，那么，只需要在程序中加入"#include <vector>"头文件包含语句即可。这里的 vector 就是指 C:\Program Files\Microsoft Visual Studio\VC98\Include\vector 文件。其他的 C++STL 容器、算法都是这样使用。

C++STL 泛型都定义在 std 命名空间中，所以，必须在头文件声明的最后加上一句"using namespace std;"。

另外，1.2 节也曾提到过，Visual Studio .NET 2008 的 C++头文件在 C:\Program Files\Microsoft Visual Studio 9.0\VC\Include 文件夹中。

如果使用 Visual Studio .NET 2008 的 VC++来编写 Win32 控制台程序，那么，C++STL 头文件包含语句和"using namespace std;"语句需要写在 stdafx.h 文件里，否则，会出现编译错误。

本章接下来详细介绍 C++STL 各种常用的容器、迭代器和算法。

2.2　vector 向量容器

vector 向量容器不但能像数组一样对元素进行随机访问，还能在尾部插入元素，是一种简单、高效的容器，完全可以代替数组。

值得注意的是，vector 具有内存自动管理的功能，对于元素的插入和删除，可动态调整所占的内存空间。

使用 vector 向量容器，需要头文件包含声明"#include <vector>"。vector 文件在 C:\Program Files\Microsoft Visual Studio\VC98\Include 文件夹中可以找到。

vector 容器的下标是从 0 开始计数的，也就是说，如果 vector 容器的大小是 n，那么，元素的下标是 $0\sim n-1$。对于 vector 容器的容量定义，可以事先定义一个固定大小，事后，可以随时调整其大小；也可以事先不定义，随时使用 push_back()方法从尾部扩张元素，也可以使用 insert()在某个元素位置前插入新元素。

vector 容器有两个重要的方法，begin()和 end()。begin()返回的是首元素位置的迭代器；end()返回的是最后一个元素的下一元素位置的迭代器。

2.2.1 创建 vector 对象

创建 vector 对象常用的有三种形式。

（1）不指定容器的元素个数，如定义一个用来存储整型的容器：

```
vector<int> v;
```

（2）创建时，指定容器的大小，如定义一个用来存储 10 个 double 类型元素的向量容器：

```
vector<double> v(10);
```

注意，元素的下标为 0～9；另外，每个元素的值被初始化为 0.0。

（3）创建一个具有 n 个元素的向量容器对象，每个元素具有指定的初始值：

```
vector<double> v(10, 8.6);
```

上述语句定义了 v 向量容器，共有 10 个元素，每个元素的值是 8.6。

2.2.2 尾部元素扩张

通常使用 push_back()对 vector 容器在尾部追加新元素。尾部追加元素，vector 容器会自动分配新内存空间。可对空的 vector 对象扩张，也可对已有元素的 vector 对象扩张。

下面的代码将 2，7，9 三个元素从尾部添加到 v 容器中，这样，v 容器中就有三个元素，其值依次是 2，7，9。

```
#include <vector>
using namespace std;
int main(int argc, char * argv[])
{
    vector<int> v;
    v.push_back(2);
    v.push_back(7);
    v.push_back(9);
    return 0;
}
```

2.2.3　下标方式访问 vector 元素

访问或遍历 vector 对象是常要做的事情。对于 vector 对象，可以采用下标方式随意访问它的某个元素，当然，也可以以下标方式对某元素重新赋值，这点类似于数组的访问方式。

下面的代码就是采用下标方式对数组赋值，再输出元素的值 2，7，9：

```cpp
#include <vector>
#include <iostream>
using namespace std;

int main(int argc, char * argv[])
{
    vector<int> v(3);
    v[0]=2;
    v[1]=7;
    v[2]=9;
    cout<<v[0]<<" "<<v[1]<<" "<<v[2]<<endl;
    return 0;
}
```

2.2.4　用迭代器访问 vector 元素

常使用迭代器配合循环语句来对 vector 对象进行遍历访问，迭代器的类型一定要与它要遍历的 vector 对象的元素类型一致。

下面的代码采用迭代器对 vector 进行了遍历，输出 2，7，9：

```cpp
#include <vector>
#include <iostream>
using namespace std;

int main(int argc, char * argv[])
{
    vector<int> v(3);
    v[0]=2;
    v[1]=7;
    v[2]=9;
    //定义迭代器变量
    vector<int>::iterator it;
    for(it=v.begin();it!=v.end();it++)
    {
        //输出迭代器上的元素值
        cout<<*it<<" ";
    }
    //换行
    cout<<endl;
    return 0;
}
```

2.2.5 元素的插入

insert()方法可以在 vector 对象的任意位置前插入一个新的元素,同时,vector 自动扩张一个元素空间,插入位置后的所有元素依次向后挪动一个位置。

要注意的是,insert()方法要求插入的位置,是元素的迭代器位置,而不是元素的下标。

下面的代码输出的结果是 8,2,1,7,9,3:

```cpp
#include <vector>
#include <iostream>
using namespace std;

int main(int argc, char * argv[])
{
    vector<int> v(3);
    v[0]=2;
    v[1]=7;
    v[2]=9;
    //在最前面插入新元素,元素值为 8
    v.insert(v.begin(),8);
    //在第 2 个元素前插入新元素 1
    v.insert(v.begin()+2,1);
    //在向量末尾追加新元素 3
    v.insert(v.end(),3);
    //定义迭代器变量
    vector<int>::iterator it;
    for(it=v.begin();it!=v.end();it++)
    {
        //输出迭代器上的元素值
        cout<<*it<<" ";
    }
    //换行
    cout<<endl;
    return 0;
}
```

2.2.6 元素的删除

erase()方法可以删除 vector 中迭代器所指的一个元素或一段区间中的所有元素。

clear()方法则一次性删除 vector 中的所有元素。

下面这段代码演示了 vector 元素的删除方法:

```cpp
#include <vector>
#include <iostream>
using namespace std;

int main(int argc, char * argv[])
{
    vector<int> v(10);
```

```
    //给向量赋值
    for(int i=0;i<10;i++)
    {
        v[i]=i;
    }
    //删除第 2 个元素,从 0 开始计数
    v.erase(v.begin()+2);
    //定义迭代器变量
    vector<int>::iterator it;
    for(it=v.begin();it!=v.end();it++)
    {
        //输出迭代器上的元素值
        cout<<*it<<" ";
    }
    //换行
    cout<<endl;
    //删除迭代器第 1 到第 5 区间的所有元素
    v.erase(v.begin()+1,v.begin()+5);
    for(it=v.begin();it!=v.end();it++)
    {
        //输出迭代器上的元素值
        cout<<*it<<" ";
    }
    //换行
    cout<<endl;
    //清空向量
    v.clear();
    //输出向量大小
    cout<<v.size()<<endl;
    return 0;
}
```

运行结果:

```
0 1 3 4 5 6 7 8 9
0 6 7 8 9
0
```

2.2.7 使用 reverse 反向排列算法

　　reverse 反向排列算法,需要定义头文件"#include <algorithm>",algorithm 文件位于 C:\Program Files\Microsoft Visual Studio\VC98\Include 文件夹中。
　　reverse 算法可将向量中某段迭代器区间元素反向排列,看下面这段代码:

```
#include <vector>
#include <iostream>
#include <algorithm>
using namespace std;

int main(int argc, char * argv[])
```

```
{
    vector<int> v(10);
    //给向量赋值
    for(int i=0;i<10;i++)
    {
        v[i]=i;
    }
    //反向排列向量的从首到尾间的元素
    reverse(v.begin(),v.end());
    //定义迭代器变量
    vector<int>::iterator it;
    for(it=v.begin();it!=v.end();it++)
    {
        //输出迭代器上的元素值
        cout<<*it<<" ";
    }
    //换行
    cout<<endl;
    return 0;
}
```

输出结果:

9 8 7 6 5 4 3 2 1 0

2.2.8 使用 sort 算法对向量元素排序

使用 sort 算法,需要声明头文件 "#include <algorithm>"。

sort 算法要求使用随机访问迭代器进行排序,在默认的情况下,对向量元素进行升序排列,下面这个程序很好地说明了 sort 算法的使用方法:

```
#include <vector>
#include <iostream>
#include <algorithm>
using namespace std;
int main(int argc, char * argv[])
{
    vector<int> v;
    int i;
    //赋值
    for(i=0;i<10;i++)
    {
        v.push_back(9-i);
    }
    //输出排序前的元素值
    for(i=0;i<10;i++)
    {
        cout<<v[i]<<" ";
    }
```

```
    //回车换行
    cout<<endl;
    //排序,升序排列
    sort(v.begin(),v.end());
    //输出排序后的元素值
    for(i=0;i<10;i++)
    {
        cout<<v[i]<<" ";
    }
    //回车换行
    cout<<endl;
    return 0;
}
```

运行结果:

9 8 7 6 5 4 3 2 1 0
0 1 2 3 4 5 6 7 8 9

还可以自己设计排序比较函数,然后,把这个函数指定给 sort 算法,那么,sort 就根据这个比较函数指定的排序规则进行排序。下面的程序自己设计了一个排序比较函数 Comp,要求对元素的值由大到小排序:

```
#include <vector>
#include <iostream>
#include <algorithm>
using namespace std;
//自己设计排序比较函数:对元素的值进行降序排列
bool Comp(const int &a,const int &b)
{
    if(a!=b)return a>b;
    else return a>b;
}
int main(int argc, char * argv[])
{
    vector<int> v;
    int i;
    //赋值
    for(i=0;i<10;i++)
    {
        v.push_back(i);
    }
    //输出排序前的元素值
    for(i=0;i<10;i++)
    {
        cout<<v[i]<<" ";
    }
    //回车换行
    cout<<endl;
    //按 Comp 函数比较规则排序
```

```
sort(v.begin(),v.end(),Comp);
//输出排序后的元素值
for(i=0;i<10;i++)
{
    cout<<v[i]<<" ";
}
//回车换行
cout<<endl;
return 0;
}
```

运行结果：

0 1 2 3 4 5 6 7 8 9
9 8 7 6 5 4 3 2 1 0

2.2.9 向量的大小

使用 size()方法可以返回向量的大小，即元素的个数。

使用 empty()方法返回向量是否为空。

下面这段代码演示了 size()方法和 empty()方法的用法：

```
#include <vector>
#include <iostream>
#include <algorithm>
using namespace std;
int main(int argc, char * argv[])
{
    vector<int> v(10);
    //给向量赋值
    for(int i=0;i<10;i++)
    {
        v[i]=i;
    }
    //输出向量的大小，即包含了多少个元素
    cout<<v.size()<<endl;
    //输出向量是否为空，如果非空，则返回逻辑假，即 0，否则返回逻辑真，即 1
    cout<<v.empty()<<endl;
    //清空向量
    v.clear();
    //输出向量是否为空，如果非空，则返回逻辑假，即 0，否则返回逻辑真，即 1
    cout<<v.empty()<<endl;
    return 0;
}
```

运行结果：

10
0
1

向量的使用方法还有很多，这里举出的只是常用的方法，如果需要更为翔实的资料，请大家参考 C++STL 相关材料。

另外，向量的元素类型可以是 int，double，char 等简单类型，也可以是结构体或 string 基本字符序列容器，使用起来非常灵活。

2.3 string 基本字符系列容器

C 语言只提供了一个 char 类型用来处理字符，而对于字符串，只能通过字符串数组来处理，显得十分不便。C++STL 提供了 string 基本字符系列容器来处理字符串，可以把 string 理解为字符串类，它提供了添加、删除、替换、查找和比较等丰富的方法。

虽然使用 vector<char>这样的向量也可以处理字符串，但功能比不上 string。向量的元素类型可以是 string，如 vector<string>这样的向量，实际上就类似于 C 语言中的字符串数组。

使用 string 容器，需要头文件包含声明"#include <string>"。string 文件在 C:\Program Files\Microsoft Visual Studio\VC98\Include 文件夹中可以找到。

2.3.1 创建 string 对象

下面这条语句创建了字符串对象 s，s 是一个空字符串，其长度为 0：

```
#include <string>
#include <iostream>
using namespace std;

int main(int argc, char * argv[])
{
    string s;
    cout<<s.length()<<endl;
    return 0;
}
```

运行结果：

0

2.3.2 给 string 对象赋值

给 string 对象赋值一般有两种方式。
（1）直接给字符串对象赋值，如：

```
#include <string>
#include <iostream>
using namespace std;

int main(int argc, char * argv[])
{
```

```
    string s;
    s="hello,C++STL.";
    cout<<s<<endl;
    return 0;
}
```

运行结果：

```
hello,C++STL.
```

（2）更常用的方法是，把字符指针赋给一个字符串对象：

```
#include <string>
#include <iostream>
using namespace std;

int main(int argc, char * argv[])
{
    string s;
    char ss[5000];
    //scanf 的输入速度比 cin 快得多
    //scanf 是 C 语言的函数，不支持 string 对象
    scanf("%s",&ss);
    //把整个字符数组赋值给 string 对象
    s=ss;
    //输出字符对象
    cout<<s<<endl;
    return 0;
}
```

运行结果（先从键盘上输入"hello,string."）：

```
hello,string.
hello,string.
```

2.3.3　从 string 对象尾部添加字符

在 string 对象的尾部添加一个字符（char），采用"+"操作符即可，具体应用如下：

```
#include <string>
#include <iostream>
using namespace std;

int main(int argc, char * argv[])
{
    string s;
    s=s+'a';
    s=s+'b';
    s=s+'c';
    cout<<s<<endl;
    return 0;
}
```

运行结果：

abc

2.3.4　从 string 对象尾部追加字符串

从尾部追加的方式有两种。

（1）直接采用"+"操作符，代码如下：

```
#include <string>
#include <iostream>
using namespace std;
int main(int argc, char * argv[])
{
    string s;
    s=s+"abc";
    s=s+"123";
    cout<<s<<endl;
    return 0;
}
```

运行结果：

abc123

（2）采用 append()方法，代码如下：

```
#include <string>
#include <iostream>
using namespace std;
int main(int argc, char * argv[])
{
    string s;
    s.append("abc");
    s.append("123");
    cout<<s<<endl;
    return 0;
}
```

运行结果：

abc123

2.3.5　给 string 对象插入字符

可以使用 insert()方法把一个字符插入到迭代器位置之前，代码如下：

```
#include <string>
#include <iostream>
using namespace std;
```

```cpp
int main(int argc, char * argv[])
{
    string s;
    s="123456";
    //定义迭代器
    string::iterator it;
    //迭代器位置为字符串首
    it=s.begin();
    //把字符'p'插入到第1个字符前(注意,字符位置是从0开始计数)
    s.insert(it+1,'p');
    cout<<s<<endl;
    return 0;
}
```

运行结果:

1p23456

2.3.6 访问 string 对象的元素

一般使用下标方式随机访问 string 对象的元素,下标是从 0 开始计数的。另外,string 对象的元素是一个字符(char),这点一定要清楚。代码如下:

```cpp
#include <string>
#include <iostream>
using namespace std;
int main(int argc, char * argv[])
{
    string s;
    s="abc123456";
    //输出 string 对象的首元素
    cout<<s[0]<<endl;
    //两个相同的字符相减值为 0
    cout<<s[0]-'a'<<endl;
    return 0;
}
```

运行结果:

a
0

2.3.7 删除 string 对象的元素

(1) 清空一个字符串,则直接给它赋空字符串即可。
(2) 使用 erase() 方法删除迭代器所指的那个元素或一个区间中的所有元素。代码如下:

```cpp
#include <string>
#include <iostream>
using namespace std;
```

```
int main(int argc, char * argv[])
{
    string s;
    s="abc123456";
    //定义迭代器变量，指向字符串对象首元素
    string::iterator it=s.begin();
    //删除第 3 个元素，元素位置从 0 开始计数
    s.erase(it+3);
    cout<<s<<endl;
    //删除 0~4 区间的所有元素
    s.erase(it,it+4);
    cout<<s<<endl;
    //清空字符串
    s="";
    //输出字符串的长度
    cout<<s.length()<<endl;
    return 0;
}
```

运行结果：

abc23456
3456
0

2.3.8 返回 string 对象的长度

采用 length()方法可返回字符串的长度；采用 empty()方法，可返回字符串是否为空，如果字符串为空，则返回逻辑真，即 1，否则，返回逻辑假，即 0。代码如下：

```
#include <string>
#include <iostream>
using namespace std;
int main(int argc, char * argv[])
{
    string s;
    s="abc123456";
    //输出字符串的长度
    cout<<s.length()<<endl;
    //清空字符串
    s="";
    //判断字符串是否为空
    cout<<s.empty()<<endl;
    return 0;
}
```

运行结果：

9
1

2.3.9 替换 string 对象的字符

使用 replace()方法可以很方便地替换 string 对象中的字符，replace()方法的重载函数相当多，常用的只有一两个，具体代码如下：

```cpp
#include <string>
#include <iostream>
using namespace std;
int main(int argc, char * argv[])
{
    string s;
    s="abc123456";
    //从第 3 个开始，将连续的 3 个字符替换为"good"
    //即将"abc"替换为"good"
    s.replace(3,3,"good");
    cout<<s<<endl;
    return 0;
}
```

运行结果：

abcgood456

2.3.10 搜索 string 对象的元素或子串

采用 find()方法可查找字符串中的第一个字符元素（char，用单引号界定）或者子串（用双引号界定），如果查到，则返回下标值（从 0 开始计数），如果查不到，则返回 4294967295。find()方法有很多重载函数，下面的代码，仅举出了一种用法。

```cpp
#include <string>
#include <iostream>
using namespace std;
int main(int argc, char * argv[])
{
    string s;
    s="cat dog cat";
    //查找第一个字符'c'，返回下标值
    cout<<s.find('c')<<endl;
    //查找第一个子串"c"，返回下标值
    cout<<s.find("c")<<endl;
    //查找第一个子串"cat"，返回下标值
    cout<<s.find("cat")<<endl;
    //查找第一个子串"dog"，返回下标值
    cout<<s.find("dog")<<endl;
    //查找第一个子串"dogc"，查不到则返回 4294967295
    cout<<s.find("dogc")<<endl;
    return 0;
}
```

运行结果：

0
0
0
4
4294967295

2.3.11　string 对象的比较

string 对象可与使用 compare()方法与其他字符串相比较。如果它比对方大，则返回 1；如果它比对方小，则返回-1；如果它与对方相同（相等），则返回 0。代码如下：

```cpp
#include <string>
#include <iostream>
using namespace std;
int main(int argc, char * argv[])
{
    string s;
    s="cat dog cat";
    //s 比"cat"字符串大，返回 1
    cout<<s.compare("cat")<<endl;
    //s 与"cat dog cat"相等，返回 0
    cout<<s.compare("cat dog cat")<<endl;
    //s 比"dog"小，返回-1
    cout<<s.compare("dog")<<endl;
    return 0;
}
```

运行结果：

1
0
-1

2.3.12　用 reverse 反向排序 string 对象

采用 reverse()方法可将 string 对象迭代器所指向的一段区间中的元素（字符）反向排序。reverse()方法需要声明头文件"#include <algorithm>"。代码如下：

```cpp
#include <string>
#include <iostream>
#include <algorithm>
using namespace std;
int main(int argc, char * argv[])
{
    string s;
    s="123456789";
    reverse(s.begin(),s.end());
```

```
    cout<<s<<endl;
    return 0;
}
```

运行结果：

```
987654321
```

2.3.13 string 对象作为 vector 元素

string 对象可以作为 vector 向量的元素，这种用法，类似于字符串数组。代码如下：

```
#include <vector>
#include <string>
#include <iostream>
#include <algorithm>
using namespace std;
int main(int argc, char * argv[])
{
    vector<string> v;
    v.push_back("Jack");
    v.push_back("Mike");
    v.push_back("Tom");
    cout<<v[0]<<endl;
    cout<<v[1]<<endl;
    cout<<v[2]<<endl;
    cout<<v[0][0]<<endl;
    cout<<v[1][0]<<endl;
    cout<<v[2].length()<<endl;
    return 0;
}
```

运行结果：

```
Jack
Mike
Tom
J
M
3
```

2.3.14 string 类型的数字化处理

在 ACM 竞赛中，常常需要将读入的数字的每位分离出来，如果采用取余的方法，花费的时间就会太长，这时候，我们可以将读入的数据当成字符串来处理，这样就方便、省时多了。下面这个程序演示了求一个整数各位的和：

```
#include <string>
#include <iostream>
using namespace std;
```

```cpp
int main(int argc, char * argv[])
{
    string s;
    s="1234059";
    int i;
    int sum=0;
    for(i=0;i<s.length();i++)
    {
        if(s[i]=='0')sum+=0;
        else if(s[i]=='1')sum+=1;
        else if(s[i]=='2')sum+=2;
        else if(s[i]=='3')sum+=3;
        else if(s[i]=='4')sum+=4;
        else if(s[i]=='5')sum+=5;
        else if(s[i]=='6')sum+=6;
        else if(s[i]=='7')sum+=7;
        else if(s[i]=='8')sum+=8;
        else if(s[i]=='9')sum+=9;
    }
    cout<<sum<<endl;
    return 0;
}
```

运行结果：

```
24
```

2.3.15 string 对象与字符数组互操作

下面这个程序演示了字符数组与 string 对象的输入与输出：

```cpp
#include <string>
#include <iostream>
using namespace std;

int main(int argc, char * argv[])
{
    string s;
    char ss[100];
    //输入字符串到字符数组中
    scanf("%s",&ss);
    //字符数组赋值线字符串对象
    s=ss;
    //用printf输出字符串对象，要采用c_str()方法
    printf(s.c_str());
    //换行
    cout<<endl;
    //用printf输出字符数组
    printf("%s",ss);
    //换行
```

```
    cout<<endl;
    //用 cout 输出字符串对象
    cout<<s<<endl;
    //用 cout 输出字符数组
    cout<<ss<<endl;
    return 0;
}
```

输出结果（从键盘输入"abc123"字符串后按回车键）：

```
abc123
abc123
abc123
abc123
abc123
```

2.3.16 string 对象与 sscanf 函数

在 C 语言中，sscanf 函数很管用，它可以把一个字符串按你需要的方式分离出子串，甚至是数字。下面这个程序演示了 sscanf 函数的具体用法：

```
#include <string>
#include <iostream>
using namespace std;

int main(int argc, char * argv[])
{
    string s1,s2,s3;
    char sa[100],sb[100],sc[100];
    //将字符串分离成子串，分隔符为空格
    sscanf("abc 123 pc","%s %s %s",sa,sb,sc);
    s1=sa;
    s2=sb;
    s3=sc;
    cout<<s1<<" "<<s2<<" "<<s3<<endl;
    //将字符串分离成数字，分隔符为空格
    //当用到数字的时候，跟 scanf 一样，它要传指针地址
    int a,b,c;
    sscanf("1 2 3","%d %d %d",&a,&b,&c);
    cout<<a<<" "<<b<<" "<<c<<endl;
    //将字符串分离成数字，分隔符为","和"$"
    //当用到数字的时候，跟 scanf 一样，它要传指针地址
    int x,y,z;
    sscanf("4,5$6","%d,%d$%d",&x,&y,&z);
    cout<<x<<" "<<y<<" "<<z<<endl;
    return 0;
}
```

运行结果：

```
abc 123 pc
```

```
1 2 3
4 5 6
```

2.3.17　string 对象与数值相互转换

有时候，string 对象与数值之间需要相互转换，下面这个例子详细说明了如何完成这项工作：

```cpp
#include <iostream>
#include <string>
#include <sstream>
using namespace std;
//C++方法：将数值转换为 string
string convertToString(double x)
 {
   ostringstream o;
   if (o << x)
      return o.str();
   return "conversion error";//if error
 }

//C++方法：将 string 转换为数值
double convertFromString(const string &s)
 {
   istringstream i(s);
   double x;
   if (i >> x)
     return x;
   return 0.0;//if error
 }

int main(int argc, char * argv[])
{
   //将数值转换为 string 的第一种方法：C 方法
   char b[10];
   string a;
   sprintf(b,"%d",1975);
   a=b;
   cout<<a<<endl;

   //将数值转换为 string 的第二种方法：C++方法
   string cc=convertToString(1976);
   cout<<cc<<endl;

   //将 string 转换为数值的方法：C++方法
   string dd="2006";
   int p=convertFromString(dd)+2;
   cout<<p<<endl;

   return 0;
 }
```

2.4 set 集合容器

　　set 集合容器实现了红黑树（Red-Black Tree）的平衡二叉检索树的数据结构，在插入元素时，它会自动调整二叉树的排列，把该元素放到适当的位置，以确保每个子树根节点的键值大于左子树所有节点的键值，而小于右子树所有节点的键值；另外，还得确保根节点左子树的高度与右子树的高度相等，这样，二叉树的高度最小，从而检索速度最快。要注意的是，它不会重复插入相同键值的元素，而采取忽略处理。图 2-3 是一个典型的红黑树。

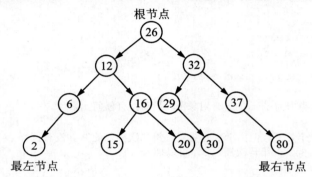

图 2-3　平衡检索二叉树（set 集合）示意图

　　平衡二叉检索树的检索使用中序遍历算法，检索效率高于 vector、deque 和 list 等容器。另外，采用中序遍历算法可将键值由小到大遍历出来，所以，可以理解为平衡二叉检索树在插入元素时，就会自动将元素按键值由小到大的顺序排列。

　　对于 set 容器中的键值，不可直接去修改。因为如果把容器中的一个键值修改了，set 容器会根据新的键值旋转子树，以保持新的平衡，这样，修改的键值很可能就不在原先那个位置上了。换句话来说，构造 set 集合的主要目的就是为了快速检索。

　　multiset（多重集合容器）、map（映照容器）和 multimap（多重映照容器）的内部结构也是平衡二叉检索树。

　　使用 set 前，需要在程序的头文件中包含声明 "#include <set>"；set 文件在 C:\Program Files\Microsoft Visual Studio\VC98\Include 文件夹中，它包含了 set 和 multiset 两种容器的定义。

2.4.1　创建 set 集合对象

　　创建 set 对象时，需要指定元素的类型，这一点与其他容器一样。下面的程序详细说明了如何创建集合对象。

```
#include <set>
using namespace std;

int main(int argc, char * argv[])
{
    //定义元素类型为 int 的集合对象 s，当前没有任何元素
    //元素的排列采用默认的比较规则，当然，可以自定义比较规则函数
```

```
    set<int> s;
    return 0;
}
```

2.4.2 元素的插入与中序遍历

采用 insert()方法把元素插入集合中去，插入的具体规则在默认的比较规则下，是按元素值由小到大插入，如果自己指定了比较规则函数，则按自定义比较规则函数插入。

使用前向迭代器对集合中序遍历，其结果正好是元素排序的结果。

下面这个例子说明了 insert()方法的使用方法：

```cpp
#include <set>
#include <iostream>
using namespace std;

int main(int argc, char * argv[])
{
    //定义元素类型为 int 的集合对象 s,当前没有任何元素
    set<int> s;
    //插入了 5 个元素，但由于 8 有重复，第二次插入的 8 并没有执行
    s.insert(8);//第一次插入 8，可以插入
    s.insert(1);
    s.insert(12);
    s.insert(6);
    s.insert(8);//第二次插入 8，重复元素，不会插入
    //中序遍历集合中的元素
    set<int>::iterator it;//定义前向迭代器
    //中序遍历集合中的所有元素
    for(it=s.begin();it!=s.end();it++)
    {
        cout<<*it<<" ";
    }
    cout<<endl;//回车换行
    return 0;
}
```

运行结果：

1 6 8 12

2.4.3 元素的反向遍历

使用反向迭代器 reverse_iterator 可以反向遍历集合，输出的结果正好是集合元素的反向排序结果。它需要用到 rbegin()和 rend()两个方法，它们分别给出了反向遍历的开始位置和结束位置。

下面这个例子详细地说明了如何对集合进行反向遍历：

```cpp
#include <set>
#include <iostream>
```

```
using namespace std;
int main(int argc, char * argv[])
{
    //定义元素类型为 int 的集合对象 s,当前没有任何元素
    set<int> s;
    //插入了 5 个元素,但由于 8 有重复,第二次插入的 8 并没有执行
    s.insert(8);//第一次插入 8,可以插入
    s.insert(1);
    s.insert(12);
    s.insert(6);
    s.insert(8);//第二次插入 8,重复元素,不会插入
    //反向遍历集合中的元素
    set<int>::reverse_iterator rit;//定义反向迭代器
    for(rit=s.rbegin();rit!=s.rend();rit++)
    {
        cout<<*rit<<" ";
    }
    cout<<endl;//回车换行
    return 0;
}
```

运行结果:

12 8 6 1

2.4.4 元素的删除

与插入元素的处理一样,集合具有高效的删除处理功能,并自动重新调整内部的红黑树的平衡。

删除的对象可以是某个迭代器位置上的元素、等于某键值的元素、一个区间上的元素和清空集合。

```
#include <set>
#include <iostream>
using namespace std;

int main(int argc, char * argv[])
{
    //定义元素类型为 int 的集合对象 s,当前没有任何元素
    set<int> s;
    //插入了 5 个元素,但由于 8 有重复,第二次插入的 8 并没有执行
    s.insert(8);//第一次插入 8,可以插入
    s.insert(1);
    s.insert(12);
    s.insert(6);
    s.insert(8);//第二次插入 8,重复元素,不会插入
    //删除键值为 6 的那个元素
    s.erase(6);
    //反向遍历集合中的元素
```

```
    set<int>::reverse_iterator rit;//定义反向迭代器
    for(rit=s.rbegin();rit!=s.rend();rit++)
    {
        cout<<*rit<<" ";
    }
    cout<<endl;//回车换行
    //清空集合
    s.clear();
    //输出集合的大小，为 0
    cout<<s.size()<<endl;
    return 0;
}
```

运行结果：

```
12 8 1
0
```

2.4.5 元素的检索

使用 find()方法对集合进行搜索，如果找到查找的键值，则返回该键值的迭代器位置，否则，返回集合最后一个元素后面的一个位置，即 end()。

下面这个程序详细讲述了如何使用 find()方法对集合进行检索：

```
#include <set>
#include <iostream>
using namespace std;

int main(int argc, char * argv[])
{
    //定义元素类型为 int 的集合对象 s，当前没有任何元素
    set<int> s;
    //插入了 5 个元素，但由于 8 有重复，第二次插入的 8 并没有执行
    s.insert(8);//第一次插入 8，可以插入
    s.insert(1);
    s.insert(12);
    s.insert(6);
    s.insert(8);//第二次插入 8，重复元素，不会插入
    set<int>::iterator it;//定义前向迭代器
    //查找键值为 6 的元素
    it=s.find(6);
    if(it!=s.end())//找到
        cout<<*it<<endl;
    else//没找到
        cout<<"not find it"<<endl;
    //查找键值为 20 的元素
    it=s.find(20);
    if(it!=s.end())//找到
        cout<<*it<<endl;
    else//没找到
```

```
        cout<<"not find it"<<endl;
    return 0;
}
```

运行结果：

```
6
not find it
```

2.4.6 自定义比较函数

使用 insert()将元素插入到集合中去的时候，集合会根据设定的比较函数将该元素放到该放的节点上去。在定义集合的时候，如果没有指定比较函数，那么采用默认的比较函数，即按键值由小到大的顺序插入元素。在很多情况下，需要自己编写比较函数。

编写比较函数有两种方法。

（1）如果元素不是结构体，那么，可以编写比较函数。下面这个程序编写的比较规则是要求按键值由大到小的顺序将元素插入到集合中：

```
#include <set>
#include <iostream>
using namespace std;
//自定义比较函数myComp, 重载"()"操作符
struct myComp
{
    bool operator()(const int &a,const int &b)
    {
        if(a!=b)
            return a>b;
        else
            return a>b;
    }
};
int main(int argc, char * argv[])
{
    //定义元素类型为 int 的集合对象 s，当前没有任何元素
    //采用的比较函数是 myComp
    set<int,myComp> s;
    //插入了 5 个元素，但由于 8 有重复，所以第二次插入的 8 并没有执行
    s.insert(8);//第一次插入 8，可以插入
    s.insert(1);
    s.insert(12);
    s.insert(6);
    s.insert(8);//第二次插入 8，重复元素，不会插入
    set<int,myComp>::iterator it;//定义前向迭代器
    for(it=s.begin();it!=s.end();it++)
    {
        cout<<*it<<" ";
    }
```

```
    cout<<endl;
    return 0;
}
```

运行结果：

12 8 6 1

（2）如果元素是结构体，那么，可以直接把比较函数写在结构体内。下面的程序详细说明了如何操作：

```
#include <set>
#include <string>
#include <iostream>
using namespace std;
struct Info
{
    string name;
    float score;
    //重载"<"操作符，自定义排序规则
    bool operator < (const Info &a) const
    {
        //按score由大到小排列。如果要由小到大排列，使用">"号即可。
        return a.score<score;
    }
};
int main(int argc, char * argv[])
{
    //定义元素类型为Info结构体的集合对象s，当前没有任何元素
    set<Info> s;
    //定义Info类型的元素
    Info info;
    //插入3个元素
    info.name="Jack";
    info.score=80.5;
    s.insert(info);
    info.name="Tomi";
    info.score=20.5;
    s.insert(info);
    info.name="Nacy";
    info.score=60.5;
    s.insert(info);
    set<Info>::iterator it;//定义前向迭代器
    for(it=s.begin();it!=s.end();it++)
    {
        cout<<(*it).name<<" : "<<(*it).score<<endl;
    }
    return 0;
}
```

运行结果：

```
Jack : 80.5
Nacy : 60.5
Tomi : 20.5
```

2.5　multiset 多重集合容器

multiset 与 set 一样，也是使用红黑树来组织元素数据的，唯一不同的是，multiset 允许重复的元素键值插入，而 set 则不允许。图 2-4 是 multiset 容器内部结构示意图。

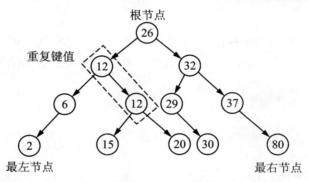

图 2-4　multiset 容器内部结构示意图

multiset 也需声明头文件包含"#include <set>"，由于它包含重复元素，所以，在插入元素、删除元素、查找元素上较 set 有差别。

2.5.1　multiset 元素的插入

下面这个程序插入了重复键值"123"，最后中序遍历了 multiset 对象。

```
#include <set>
#include <string>
#include <iostream>
using namespace std;
int main(int argc, char * argv[])
{
    //定义元素类型为 string 的多重集合对象 s，当前没有任何元素
    multiset<string> ms;
    ms.insert("abc");
    ms.insert("123");
    ms.insert("111");
    ms.insert("aaa");
    ms.insert("123");
    multiset<string>::iterator it;
    for(it=ms.begin();it!=ms.end();it++)
    {
```

```
        cout<<*it<<endl;
    }
    return 0;
}
```

运行结果:

111
123
123
aaa
abc

2.5.2 multiset 元素的删除

采用 erase()方法可以删除 multiset 对象中的某个迭代器位置上的元素、某段迭代器区间中的元素、键值等于某个值的所有重复元素,并返回删除元素的个数。采用 clear()方法可以清空元素。

下面这个程序详细说明了 insert()方法的使用方法。

```
#include <set>
#include <string>
#include <iostream>
using namespace std;
int main(int argc, char * argv[])
{
    //定义元素类型为 string 的多重集合对象 s,当前没有任何元素
    multiset<string> ms;
    ms.insert("abc");
    ms.insert("123");
    ms.insert("111");
    ms.insert("aaa");
    ms.insert("123");
    multiset<string>::iterator it;
    for(it=ms.begin();it!=ms.end();it++)
    {
        cout<<*it<<endl;
    }
    //删除值为"123"的所有重复元素,返回删除元素总数 2
    int n=ms.erase("123");
    cout<<"Total deleted : "<<n<<endl;
    //输出删除后的剩余元素
    cout<<"all elements after deleted :"<<endl;
    for(it=ms.begin();it!=ms.end();it++)
    {
        cout<<*it<<endl;
    }
    return 0;
}
```

运行结果:
```
111
123
123
aaa
abc
Total deleted : 2
all elements after deleted :
111
aaa
abc
```

2.5.3 查找元素

使用 find()方法查找元素,如果找到,则返回该元素的迭代器位置(如果该元素存在重复,则返回第一个元素重复元素的迭代器位置);如果没有找到,则返回 end()迭代器位置。

下面的程序具体说明了 find()方法的作用方法。

```cpp
#include <set>
#include <string>
#include <iostream>
using namespace std;

int main(int argc, char * argv[])
{
    //定义元素类型为string的多重集合对象s,当前没有任何元素
    multiset<string> ms;
    ms.insert("abc");
    ms.insert("123");
    ms.insert("111");
    ms.insert("aaa");
    ms.insert("123");
    multiset<string>::iterator it;
    //查找键值"123"
    it=ms.find("123");
    if(it!=ms.end())//找到
    {
        cout<<*it<<endl;
    }
    else//没有找到
    {
        cout<<"not find it"<<endl;
    }
    it=ms.find("bbb");
    if(it!=ms.end())//找到
    {
        cout<<*it<<endl;
    }
```

```
        else//没有找到
        {
            cout<<"not find it"<<endl;
        }
        return 0;
}
```

运行结果：

```
123
not find it
```

2.6 map 映照容器

map 映照容器的元素数据是由一个键值和一个映照数据组成的，键值与映照数据之间具有一一映照的关系。

map 映照容器的数据结构也是采用红黑树来实现的，插入元素的键值不允许重复，比较函数只对元素的键值进行比较，元素的各项数据可通过键值检索出来。由于 map 与 set 采用的都是红黑树的数据结构，所以，用法基本相似。图 2-5 是 map 映照容器元素的数据构成示意图。

键值	映照数据
Name	Score
Jack	98.5
Bomi	96.0
Kate	97.5

图 2-5 map 映照容器元素的数据构成示意图

使用 map 容器需要头文件包含语句"#include <map>"，map 文件在 C:\Program Files\Microsoft Visual Studio\VC98\Include 文件夹内。map 文件也包含了对 multimap 多重映照容器的定义。

2.6.1 map 创建、元素插入和遍历访问

创建 map 对象，键值与映照数据的类型由自己定义。在没有指定比较函数时，元素的插入位置是按键值由小到大插入到黑白树中去的，这点和 set 一样。下面这个程序详细说明了如何操作 map 容器。

```
#include <map>
#include <string>
#include <iostream>
using namespace std;
int main(int argc, char * argv[])
{
```

```
    //定义map对象，当前没有任何元素
    map<string,float> m;
    //插入元素，按键值的由小到大放入黑白树中
    m["Jack"]=98.5;
    m["Bomi"]=96.0;
    m["Kate"]=97.5;
    //前向遍历元素
    map<string,float>::iterator it;
    for(it=m.begin();it!=m.end();it++)
    {
        //输出键值与映照数据
        cout<<(*it).first<<" : "<<(*it).second<<endl;
    }
    return 0;
}
```

运行结果：

```
Bomi : 96
Jack : 98.5
Kate : 97.5
```

程序编译时，会产生代号为"warning C4786"的警告，"4786"是标记符超长警告的代号。可以在程序的头文件包含代码的前面使用"#pragma warning(disable:4786)"宏语句，强制编译器忽略该警告。4786号警告对程序的正确性和运行并无影响。

2.6.2 删除元素

与 set 容器一样，map 映照容器的 erase()删除元素函数，可以删除某个迭代器位置上的元素、等于某个键值的元素、一个迭代器区间上的所有元素，当然，也可使用 clear()方法清空 map 映照容器。

下面这个程序演示了删除 map 容器中键值为 28 的元素：

```
#pragma warning(disable:4786)
#include <map>
#include <string>
#include <iostream>
using namespace std;
int main(int argc, char * argv[])
{
    //定义map对象，当前没有任何元素
    map<int,char> m;
    //插入元素，按键值的由小到大放入黑白树中
    m[25]='m';
    m[28]='k';
    m[10]='x';
    m[30]='a';
    //删除键值为28的元素
    m.erase(28);
```

```
        //前向遍历元素
        map<int,char>::iterator it;
        for(it=m.begin();it!=m.end();it++)
        {
            //输出键值与映照数据
            cout<<(*it).first<<" : "<<(*it).second<<endl;
        }
        return 0;
    }
```

运行结果：

```
10 : x
25 : m
30 : a
```

2.6.3 元素反向遍历

可以使用反向迭代器 reverse_iterator 反向遍历 map 照映容器中的数据，它需要 rbegin() 方法和 rend()方法指出反向遍历的起始位置和终止位置。

```
#pragma warning(disable:4786)
#include <map>
#include <string>
#include <iostream>
using namespace std;

int main(int argc, char * argv[])
{
    //定义map对象，当前没有任何元素
    map<int,char> m;
    //插入元素，按键值的由小到大放入黑白树中
    m[25]='m';
    m[28]='k';
    m[10]='x';
    m[30]='a';
    //反向遍历元素
    map<int,char>::reverse_iterator rit;
    for(rit=m.rbegin();rit!=m.rend();rit++)
    {
        //输出键值与映照数据
        cout<<(*rit).first<<" : "<<(*rit).second<<endl;
    }
    return 0;
}
```

运行结果：

```
30 : a
28 : k
```

```
25 : m
10 : x
```

2.6.4 元素的搜索

使用 find()方法来搜索某个键值,如果搜索到了,则返回该键值所在的迭代器位置,否则,返回 end()迭代器位置。由于 map 采用黑白树数据结构来实现,所以搜索速度是极快的。

下面这个程序搜索键值为 28 的元素:

```
#pragma warning(disable:4786)
#include <map>
#include <string>
#include <iostream>
using namespace std;

int main(int argc, char * argv[])
{
    //定义map对象,当前没有任何元素
    map<int,char> m;
    //插入元素,按键值的由小到大放入黑白树中
    m[25]='m';
    m[28]='k';
    m[10]='x';
    m[30]='a';
    map<int,char>::iterator it;
    it=m.find(28);
    if(it!=m.end())//搜索到该键值
    {
        cout<<(*it).first<<" : "<<(*it).second<<endl;
    }
    else
    {
        cout<<"not found it"<<endl;
    }
    return 0;
}
```

运行结果:

```
28 : k
```

2.6.5 自定义比较函数

将元素插入到 map 中去的时候,map 会根据设定的比较函数将该元素放到该放的节点上去。在定义 map 的时候,如果没有指定比较函数,那么采用默认的比较函数,即按键值由小到大的顺序插入元素。在很多情况下,需要自己编写比较函数。

编写比较函数与 set 比较函数是一致的,因为它们的内部数据结构都是红黑树。编写方法有两种。

(1) 如果元素不是结构体,那么,可以编写比较函数。下面这个程序编写的比较规则是要求按键值由大到小的顺序将元素插入到 map 中:

```
#pragma warning(disable:4786)
#include <map>
#include <string>
#include <iostream>
using namespace std;
//自定义比较函数 myComp
struct myComp
{
    bool operator()(const int &a,const int &b)
    {
        if(a!=b)return a>b;
        else
            return a>b;
    }
};
int main(int argc, char * argv[])
{
    //定义 map 对象,当前没有任何元素
    map<int,char,myComp> m;
    //插入元素,按键值的由小到大放入黑白树中
    m[25]='m';
    m[28]='k';
    m[10]='x';
    m[30]='a';
    //使用前向迭代器中序遍历 map
    map<int,char,myComp>::iterator it;
    for(it=m.begin();it!=m.end();it++)
    {
        cout<<(*it).first<<" : "<<(*it).second<<endl;
    }
    return 0;
}
```

运行结果:

30 : a
28 : k
25 : m
10 : x

(2) 如果元素是结构体,那么,可以直接把比较函数写在结构体内。下面的程序详细说明了如何操作:

```
#pragma warning(disable:4786)
#include <map>
```

```cpp
#include <string>
#include <iostream>
using namespace std;
struct Info
{
    string name;
    float score;
    //重载 "<" 操作符，自定义排序规则
    bool operator < (const Info &a) const
    {
        //按 score 由大到小排列。如果要由小到大排列，使用 ">" 号即可
        return a.score<score;
    }
};
int main(int argc, char * argv[])
{
    //定义 map 对象，当前没有任何元素
    map<Info,int> m;
    //定义 Info 结构体变量
    Info info;
    //插入元素，按键值的由小到大放入黑白树中
    info.name="Jack";
    info.score=60;
    m[info]=25;
    info.name="Bomi";
    info.score=80;
    m[info]=10;
    info.name="Peti";
    info.score=66.5;
    m[info]=30;
    //使用前向迭代器中序遍历 map
    map<Info,int>::iterator it;
    for(it=m.begin();it!=m.end();it++)
    {
        cout<<(*it).second<<" : ";
        cout<<((*it).first).name<<" "<<((*it).first).score<<endl;
    }
    return 0;
}
```

运行结果：

```
10 : Bomi 80
30 : Peti 66.5
25 : Jack 60
```

2.6.6 用 map 实现数字分离

对数字的各位进行分离，采用取余等数学方法操作是很耗时的。而把数字当成字符串，

使用 map 的映照功能，很方便地实现了数字分离。下面这个程序将一个字符串中的字符当成数字，并将各位的数值相加，最后输出各位的和。

```cpp
#pragma warning(disable:4786)
#include <string>
#include <map>
#include <iostream>
using namespace std;
int main(int argc, char * argv[])
{
    //定义map对象，当前没有任何元素
    map<char,int> m;
    //赋值:字符映射数字
    m['0']=0;
    m['1']=1;
    m['2']=2;
    m['3']=3;
    m['4']=4;
    m['5']=5;
    m['6']=6;
    m['7']=7;
    m['8']=8;
    m['9']=9;
    /*上面的10条赋值语句可采用下面这个循环来简化代码编写
    for(int j=0;j<10;j++)
    {
        m['0'+j]=j;
    }
    */
    string sa,sb;
    sa="6234";
    int i;
    int sum=0;
    for(i=0;i<sa.length();i++)
    {
        sum+=m[sa[i]];
    }
    cout<<"sum = "<<sum<<endl;
    return 0;
}
```

运行结果：

```
sum = 15
```

2.6.7 数字映照字符的 map 写法

在很多情况下，需要实现将数字映射为相应的字符，看看下面这个程序：

```cpp
#pragma warning(disable:4786)
```

```
#include <map>
#include <string>
#include <iostream>
using namespace std;

int main(int argc, char * argv[])
{
    //定义map对象,当前没有任何元素
    map<int,char> m;
    //赋值:字符映射数字
    m[0]='0';
    m[1]='1';
    m[2]='2';
    m[3]='3';
    m[4]='4';
    m[5]='5';
    m[6]='6';
    m[7]='7';
    m[8]='8';
    m[9]='9';
    /*上面的10条赋值语句可采用下面这个循环来简化代码编写
    for(int j=0;j<10;j++)
    {
        m[j]='0'+j;
    }
    */
    int n=7;
    string s="The number is ";
    cout<<s + m[n]<<endl;
    return 0;
}
```

运行结果:

```
The number is 7
```

2.7 multimap 多重映照容器

multimap 与 map 基本相同,唯独不同的是,multimap 允许插入重复键值的元素。由于允许重复键值存在,所以,multimap 的元素插入、删除、查找都与 map 不相同。

要使用 multimap,则需要头文件包含语句"#include <map>"。map 文件在 C:\Program Files\Microsoft Visual Studio\VC98\Include 文件夹中。

2.7.1 multimap 对象创建、元素插入

可以重复插入元素,插入元素需要使用 insert()方法。下面这个程序,重复插入了一个名为"Jack"的键值。

49

```
#pragma warning(disable:4786)
#include <map>
#include <string>
#include <iostream>
using namespace std;

int main(int argc, char * argv[])
{
    //定义 map 对象，当前没有任何元素
    multimap<string,double> m;
    //插入元素
    m.insert(pair<string,double>("Jack",300.5));
    m.insert(pair<string,double>("Kity",200));
    m.insert(pair<string,double>("Memi",500));
    //重复插入键值"Jack"
    m.insert(pair<string,double>("Jack",306));
    //使用前向迭代器中序遍历 multimap
    multimap<string,double>::iterator it;
    for(it=m.begin();it!=m.end();it++)
    {
        cout<<(*it).first<<" : "<<(*it).second<<endl;
    }
    return 0;
}
```

运行结果：

```
Jack : 300.5
Jack : 306
Kity : 200
Memi : 500
```

2.7.2 元素的删除

删除操作采用 erase()方法，可删除某个迭代器位置上的元素、等于某个键值的所有重复元素、一个迭代器区间上的元素。使用 clear()方法可将 multimap 容器中的元素清空。

因为有重复的键值，所以，删除操作会将要删除的键值一次性从 multimap 中删除。下面这个程序说明了这点。

```
#pragma warning(disable:4786)
#include <map>
#include <string>
#include <iostream>
using namespace std;

int main(int argc, char * argv[])
{
    //定义 map 对象，当前没有任何元素
    multimap<string,double> m;
    //插入元素
```

```cpp
    m.insert(pair<string,double>("Jack",300.5));
    m.insert(pair<string,double>("Kity",200));
    m.insert(pair<string,double>("Memi",500));
    //重复插入键值"Jack"
    m.insert(pair<string,double>("Jack",306));
    //使用前向迭代器中序遍历multimap
    multimap<string,double>::iterator it;
    for(it=m.begin();it!=m.end();it++)
    {
        cout<<(*it).first<<" : "<<(*it).second<<endl;
    }
    //删除键值等于"Jack"的元素
    m.erase("Jack");
    //使用前向迭代器中序遍历multimap
    cout<<"the elements after deleted:"<<endl;
    for(it=m.begin();it!=m.end();it++)
    {
        cout<<(*it).first<<" : "<<(*it).second<<endl;
    }
    return 0;
}
```

运行结果：

```
Jack : 300.5
Jack : 306
Kity : 200
Memi : 500
the elements after deleted:
Kity : 200
Memi : 500
```

2.7.3 元素的查找

由于 multimap 存在重复的键值，所以 find()方法只返回重复键值中的第一个元素的迭代器位置，如果没有找到该键值，则返回 end()迭代器位置。下面这个程序说明了这一点。

```cpp
#pragma warning(disable:4786)
#include <map>
#include <string>
#include <iostream>
using namespace std;
int main(int argc, char * argv[])
{
    //定义map对象，当前没有任何元素
    multimap<string,double> m;
    //插入元素
    m.insert(pair<string,double>("Jack",300.5));
    m.insert(pair<string,double>("Kity",200));
    m.insert(pair<string,double>("Memi",500));
```

```cpp
//重复插入键值"Jack"
m.insert(pair<string,double>("Jack",306));
//使用前向迭代器中序遍历multimap
multimap<string,double>::iterator it;
cout<<"all of the elements:"<<endl;
for(it=m.begin();it!=m.end();it++)
{
    cout<<(*it).first<<" : "<<(*it).second<<endl;
}
//查找键值
cout<<endl<<"the searching result:"<<endl;
it=m.find("Jack");
if(it!=m.end())//找到
{
    cout<<(*it).first<<" "<<(*it).second<<endl;
}
else//没找到
{
    cout<<"not find it"<<endl;
}
it=m.find("Nacy");
if(it!=m.end())//找到
{
    cout<<(*it).first<<" "<<(*it).second<<endl;
}
else//没找到
{
    cout<<"not find it"<<endl;
}
return 0;
}
```

运行结果：

```
all of the elements:
Jack : 300.5
Jack : 306
Kity : 200
Memi : 500

the searching result:
Jack 300.5
not find it
```

2.8 deque 双端队列容器

deque 双端队列容器与 vector 一样，采用线性表顺序存储结构。但与 vector 唯一不同的是，deque 采用分块的线性存储结构来存储数据，每块的大小一般为 512 字节，称为一个 deque 块，所有的 deque 块使用一个 Map 块进行管理，每个 Map 数据项记录各个 deque

块的首地址。这样一来，deque 块在头部和尾部都可插入和删除元素，而不需移动其他元素（使用 push_back()方法在尾部插入元素，会扩张队列；而使用 push_front()方法在首部插入元素和使用 insert()方法在中间插入元素，只是将原位置上的元素值覆盖，不会增加新元素）。一般来说，当考虑到容器元素的内存分配策略和操作的性能时，deque 相对于 vector 更有优势。双端队列容器结构示意图如图 2-6 所示。

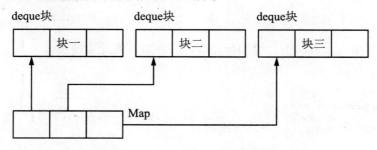

图 2-6　双端队列容器结构示意图

使用 deque 需要声明头文件包含"#include <deque>"，文件 deque 在 C:\Program Files\Microsoft Visual Studio\VC98\Include 文件夹中。

2.8.1　创建 deque 对象

创建 deque 对象的方法通常有三种。

（1）创建没有任何元素的 deque 对象，如：

```
deque<int> d;
deque<float> dd;
```

（2）创建具有 *n* 个元素的 deque 对象，如：

```
deque<int> d(10);    //创建具有 10 个整型元素的 deque 对象 d
```

（3）创建具有 *n* 个元素的 deque 对象，并赋初值，如：

```
deque<int> d(10,8.5);    //创建具有 10 个整型元素的 deque 对象 d，每个元素值为 8.5
```

2.8.2　插入元素

（1）使用 push_back()方法从尾部插入元素，会不断扩张队列。

```
#include <deque>
#include <iostream>
using namespace std;
int main(int argc, char * argv[])
{
    //定义 deque 对象，元素类型是整型
    deque<int> d;
    //从尾部连续插入三个元素
    d.push_back(1);
    d.push_back(2);
```

```
        d.push_back(3);
        //以数组方式输出元素
        cout<<d[0]<<" "<<d[1]<<" "<<d[2]<<endl;
        return 0;
    }
```

运行结果：

1 2 3

（2）从头部插入元素，不会增加新元素，只将原有的元素覆盖。

```
    #include <deque>
    #include <iostream>
    using namespace std;
    int main(int argc, char * argv[])
    {
        //定义 deque 对象，元素类型是整型
        deque<int> d;
        //从尾部连续插入三个元素
        d.push_back(1);
        d.push_back(2);
        d.push_back(3);
        //从头部插入元素，不会增加新元素，只将原有的元素覆盖
        d.push_front(10);
        d.push_front(20);
        //以数组方式输出元素
        cout<<d[0]<<" "<<d[1]<<" "<<d[2]<<endl;
        return 0;
    }
```

运行结果：

20 10 1

（3）从中间插入元素，不会增加新元素，只将原有的元素覆盖。

```
    #include <deque>
    #include <iostream>
    using namespace std;
    int main(int argc, char * argv[])
    {
        //定义 deque 对象，元素类型是整型
        deque<int> d;
        //从尾部连续插入三个元素
        d.push_back(1);
        d.push_back(2);
        d.push_back(3);
        //中间插入元素,不会增加新元素，只将原有的元素覆盖
        d.insert(d.begin()+1,88);
        //以数组方式输出元素
```

```
    cout<<d[0]<<" "<<d[1]<<" "<<d[2]<<endl;
    return 0;
}
```

运行结果：

1 88 2

2.8.3 前向遍历

（1）以数组方式遍历。

```
#include <deque>
#include <iostream>
using namespace std;

int main(int argc, char * argv[])
{
    //定义deque对象，元素类型是整型
    deque<int> d;
    //从尾部连续插入三个元素
    d.push_back(1);
    d.push_back(2);
    d.push_back(3);
    //以数组方式输出元素
    int i;
    for(i=0;i<d.size();i++)
    {
        cout<<d[i]<<" ";
    }
    //回车换行
    cout<<endl;
    return 0;
}
```

运行结果：

1 2 3

（2）以前向迭代器的方式遍历。

```
#include <deque>
#include <iostream>
using namespace std;

int main(int argc, char * argv[])
{
    //定义deque对象，元素类型是整型
    deque<int> d;
    //从尾部连续插入三个元素
    d.push_back(1);
    d.push_back(2);
    d.push_back(3);
```

```cpp
//以前向迭代器的方式遍历
deque<int>::iterator it;
for(it=d.begin();it!=d.end();it++)
{
    cout<<*it<<" ";
}
//回车换行
cout<<endl;
return 0;
}
```

运行结果:

1 2 3

2.8.4 反向遍历

采用反向迭代器对双端队列容器进行反向遍历。

```cpp
#include <deque>
#include <iostream>
using namespace std;

int main(int argc, char * argv[])
{
    //定义 deque 对象，元素类型是整型
    deque<int> d;
    //从尾部连续插入三个元素
    d.push_back(1);
    d.push_back(2);
    d.push_back(3);
    //以反向迭代器的方式遍历
    deque<int>::reverse_iterator rit;
    for(rit=d.rbegin();rit!=d.rend();rit++)
    {
        cout<<*rit<<" ";
    }
    //回车换行
    cout<<endl;
    return 0;
}
```

运行结果:

3 2 1

2.8.5 删除元素

可以从双端队列容器的首部、尾部、中部删除元素，并可以清空双端队列容器。下面分别举例说明这 4 种删除元素的操作方法。

(1) 采用 pop_front()方法从头部删除元素。

```cpp
#include <deque>
#include <iostream>
using namespace std;

int main(int argc, char * argv[])
{
    //定义 deque 对象，元素类型是整型
    deque<int> d;
    //从尾部连续插入五个元素
    d.push_back(1);
    d.push_back(2);
    d.push_back(3);
    d.push_back(4);
    d.push_back(5);
    //从头部删除元素
    d.pop_front();
    d.pop_front();
    //以前向迭代器的方式遍历
    deque<int>::iterator it;
    for(it=d.begin();it!=d.end();it++)
    {
        cout<<*it<<" ";
    }
    //回车换行
    cout<<endl;
    return 0;
}
```

运行结果：

3 4 5

(2) 采用 pop_back()方法从尾部删除元素。

```cpp
#include <deque>
#include <iostream>
using namespace std;

int main(int argc, char * argv[])
{
    //定义 deque 对象，元素类型是整型
    deque<int> d;
    //从尾部连续插入五个元素
    d.push_back(1);
    d.push_back(2);
    d.push_back(3);
    d.push_back(4);
    d.push_back(5);
    //从尾部删除元素
    d.pop_back();
```

```
    //以前向迭代器的方式遍历
    deque<int>::iterator it;
    for(it=d.begin();it!=d.end();it++)
    {
        cout<<*it<<" ";
    }
    //回车换行
    cout<<endl;
    return 0;
}
```

运行结果：

```
1 2 3 4
```

（3）使用 erase()方法从中间删除元素，其参数是迭代器位置。

```
#include <deque>
#include <iostream>
using namespace std;
int main(int argc, char * argv[])
{
    //定义 deque 对象，元素类型是整型
    deque<int> d;
    //从尾部连续插入五个元素
    d.push_back(1);
    d.push_back(2);
    d.push_back(3);
    d.push_back(4);
    d.push_back(5);
    //从中间删除元素，erase 的参数是迭代器位置
    d.erase(d.begin()+1);
    //以前向迭代器的方式遍历
    deque<int>::iterator it;
    for(it=d.begin();it!=d.end();it++)
    {
        cout<<*it<<" ";
    }
    //回车换行
    cout<<endl;
    return 0;
}
```

运行结果：

```
1 3 4 5
```

（4）使用 clear()方法清空 deque 对象。

```
#include <deque>
#include <iostream>
using namespace std;
```

```
int main(int argc, char * argv[])
{
    //定义 deque 对象，元素类型是整型
    deque<int> d;
    //从尾部连续插入五个元素
    d.push_back(1);
    d.push_back(2);
    d.push_back(3);
    d.push_back(4);
    d.push_back(5);
    //清空元素
    d.clear();
    //输出元素的个数
    cout<<d.size()<<endl;
    return 0;
}
```

运行结果：

0

2.9　list 双向链表容器

　　list 容器实现了双向链表的数据结构，数据元素是通过链表指针串连成逻辑意义上的线性表，这样，对链表的任一位置的元素进行插入、删除和查找都是极快速的。图 2-7 是 list 采用的双向循环链表的结构示意图。

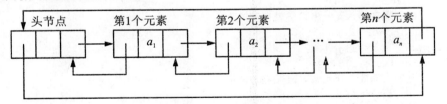

图 2-7　list 采用的双向循环链表的结构示意图

　　list 的每个节点有三个域：前驱元素指针域、数据域和后继元素指针域。前驱元素指针域保存了前驱元素的首地址；数据域则是本节点的数据；后继元素指针域则保存了后继元素的首地址。list 的头节点的前驱元素指针域保存的是链表中尾元素的首地址，而 list 的尾节点的后继元素指针域则保存了头节点的首地址，这样，就构成了一个双向循环链。

　　由于 list 对象的节点并不要求在一段连续的内存中，所以，对于迭代器，只能通过"++"或"--"的操作将迭代器移动到后继/前驱节点元素处。而不能对迭代器进行+n 或-n 的操作，这点，是与 vector 等不同的地方。

　　使用 list 需要声明头文件包含"#include <list>"，list 文件在 C:\Program Files\Microsoft Visual Studio\VC98\Include 文件夹中。

2.9.1 创建 list 对象

（1）创建空链表，如：

　`list<int> l;`

（2）创建具有 n 个元素的链表，如：

　`list<int> l(10);  //创建具有10个元素的链表`

2.9.2 元素插入和遍历

有三种方法往链表里插入新元素：
（1）采用 push_back() 方法往尾部插入新元素，链表自动扩张。
（2）采用 push_front() 方法往首部插入新元素，链表自动扩张。
（3）采用 insert() 方法往迭代器位置处插入新元素，链表自动扩张。注意，迭代器只能进行"++"或"--"操作，不能进行+n 或-n 的操作，因为元素的位置并不是物理相连的。

采用前向迭代器 iterator 对链表进行遍历。

下面的程序详细说明了如何对 list 进行元素插入和遍历。

```
#include <list>
#include <iostream>
using namespace std;

int main(int argc, char * argv[])
{
    //定义元素为整型的list对象,当前没有元素
    list<int> l;
    //在链表尾部插入新元素,链表自动扩张
    l.push_back(2);
    l.push_back(1);
    l.push_back(5);
    //在链表头部插入新元素,链表自动扩张
    l.push_front(8);
    //在任意位置插入新元素,链表自动扩张
    list<int>::iterator it;
    it=l.begin();
    it++;//注意,链表的迭代器只能进行++或--操作,而不能进行+n操作
    l.insert(it,20);
    //使用前向迭代器遍历链表
    for(it=l.begin();it!=l.end();it++)
    {
        cout<<*it<<" ";
    }
    //回车换行
    cout<<endl;
    return 0;
}
```

2.9.3 反向遍历

采用反向迭代器 reverse_iterator 对链表进行反向遍历。

```
#include <list>
#include <iostream>
4using namespace std;
int main(int argc, char * argv[])
{
    //定义元素为整型的list对象，当前没有元素
    list<int> l;
    //在链表尾部插入新元素，链表自动扩张
    l.push_back(2);
    l.push_back(1);
    l.push_back(5);
    //反向遍历链表
    list<int>::reverse_iterator rit;
    for(rit=l.rbegin();rit!=l.rend();rit++)
    {
        cout<<*rit<<" ";
    }
    //回车换行
    cout<<endl;
    return 0;
}
```

运行结果：

5 1 2

2.9.4 元素删除

（1）可以使用 remove() 方法删除链表中一个元素，值相同的元素都会被删除。

```
#include <list>
#include <iostream>
using namespace std;
int main0028int argc, char * argv[])
{
    //定定元素为整型的list对象，当前没有元素
    list<int> l;
    //在链表尾部插入新元素，链表自动扩张
    l.push_back(2);
    l.push_back(1);
    l.push_back(5);
```

```
    l.push_back(1);
    //遍历链表
    list<int>::iterator it;
    for(it=l.begin();it!=l.end();it++)
    {
        cout<<*it<<" ";
    }
    //回车换行
    cout<<endl;
    //删除值等于1的所有元素
    l.remove(1);
    //遍历链表
    for(it=l.begin();it!=l.end();it++)
    {
        cout<<*it<<" ";
    }
    //回车换行
    cout<<endl;
    return 0;
}
```

运行结果

2 1 5 1
2 5

（2）使用 pop_front()方法删除链表首元素，使用 pop_back()方法删除链表尾元素。

```
#include <list>
#include <iostream>
using namespace std;

int main(int argc, char * argv[])
{
    //定义元素为整型的list对象，当前没有元素
    list<int> l;
    //在链表尾部插入新元素，链表自动扩张
    l.push_back(2);
    l.push_back(8);
    l.push_back(1);
    l.push_back(5);
    l.push_back(1);
    //遍历链表
    list<int>::iterator it;
    for(it=l.begin();it!=l.end();it++)
    {
        cout<<*it<<" ";
    }
    //回车换行
    cout<<endl;
    //删除首元素
    l.pop_front();
```

```
    //删除尾元素
    l.pop_back();
    //遍历链表
    for(it=l.begin();it!=l.end();it++)
    {
        cout<<*it<<" ";
    }
    //回车换行
    cout<<endl;
    return 0;
}
```

运行结果：

```
2 8 1 5 1
8 1 5
```

（3）使用 erase()方法删除迭代器位置上的元素。

```
#include <list>
#include <iostream>
using namespace std;

int main(int argc, char * argv[])
{
    //定义元素为整型的 list 对象，当前没有元素
    list<int> l;
    //在链表尾部插入新元素，链表自动扩张
    l.push_back(2);
    l.push_back(8);
    l.push_back(1);
    l.push_back(5);
    l.push_back(1);
    //遍历链表
    list<int>::iterator it;
    for(it=l.begin();it!=l.end();it++)
    {
        cout<<*it<<" ";
    }
    //回车换行
    cout<<endl;
    //删除第 2 个元素（从 0 开始计数）
    it=l.begin();
    it++;
    it++;
    l.erase(it);
    //遍历链表
    for(it=l.begin();it!=l.end();it++)
    {
        cout<<*it<<" ";
    }
```

```
    //回车换行
    cout<<endl;
    return 0;
}
```

运行结果:

```
2 8 1 5 1
2 8 5 1
```

(4) 使用 clear()方法清空链表。

```
#include <list>
#include <iostream>
using namespace std;

int main(int argc, char * argv[])
{
    //定义元素为整型的list对象,当前没有元素
    list<int> l;
    //在链表尾部插入新元素,链表自动扩张
    l.push_back(2);
    l.push_back(8);
    l.push_back(1);
    l.push_back(5);
    l.push_back(1);
    //遍历链表
    list<int>::iterator it,it2;
    for(it=l.begin();it!=l.end();it++)
    {
        cout<<*it<<" ";
    }
    //回车换行
    cout<<endl;
    //清空链表
    l.clear();
    //打印链表元素个数
    cout<<l.size()<<endl;
    return 0;
}
```

运行结果:

```
2 8 1 5 1
0
```

2.9.5 元素查找

采用 find()查找算法可以在链表中查找元素,如果找到该元素,返回的是该元素的迭代器位置;如果没有找到,则返回 end()迭代器位置。

find()算法需要声明头文件包含语句"#include <algorithm>"。

```cpp
#include <list>
#include <iostream>
#include <algorithm>
using namespace std;

int main(int argc, char * argv[])
{
    //定义元素为整型的list对象,当前没有元素
    list<int> l;
    //在链表尾部插入新元素,链表自动扩张
    l.push_back(2);
    l.push_back(8);
    l.push_back(1);
    l.push_back(5);
    l.push_back(1);
    //遍历链表
    list<int>::iterator it,it2;
    for(it=l.begin();it!=l.end();it++)
    {
        cout<<*it<<" ";
    }
    //回车换行
    cout<<endl;
    //采用find()查找算法在链表中查找
    it=find(l.begin(),l.end(),5);
    if(it!=l.end())//找到
    {
        cout<<"find it"<<endl;
    }
    else
    {
        cout<<"not find it"<<endl;
    }
    it=find(l.begin(),l.end(),10);
    if(it!=l.end())//找到
    {
        cout<<"find it"<<endl;
    }
    else
    {
        cout<<"not find it"<<endl;
    }
    return 0;
}
```

运行结果:

```
2 8 1 5 1
find it
not find it
```

2.9.6 元素排序

采用 sort()方法可以对链表元素进行升序排列。

```cpp
#include <list>
#include <iostream>
using namespace std;
int main(int argc, char * argv[])
{
    //定义元素为整型的 list 对象，当前没有元素
    list<int> l;
    //在链表尾部插入新元素，链表自动扩张
    l.push_back(2);
    l.push_back(8);
    l.push_back(1);
    l.push_back(5);
    l.push_back(1);
    //遍历链表
    list<int>::iterator it,it2;
    for(it=l.begin();it!=l.end();it++)
    {
        cout<<*it<<" ";
    }
    //回车换行
    cout<<endl;
    //使用 sort()方法对链表排序，是升序排列
    l.sort();
    //遍历链表
    for(it=l.begin();it!=l.end();it++)
    {
        cout<<*it<<" ";
    }
    //回车换行
    cout<<endl;
    return 0;
}
```

运行结果：

```
2 8 1 5 1
1 1 2 5 8
```

2.9.7 剔除连续重复元素

采用 unique()方法可以剔除连续重复元素，只保留一个。

```cpp
#include <list>
#include <iostream>
using namesppa4ce std;
```

```
int main(int argc, char * argv[])
{
    //定义元素为整型的list对象，当前没有元素
    list<int> l;
    //在链表尾部插入新元素，链表自动扩张
    l.push_back(2);
    l.push_back(8);
    l.push_back(1);
    l.push_back(1);
    l.push_back(1);
    l.push_back(5);
    l.push_back(1);
    //遍历链表
    list<int>::iterator it,it2;
    for(it=l.begin();it!=l.end();it++)
    {
        cout<<*it<<" ";
    }
    //回车换行
    cout<<endl;
    //剔除连续重复元素（只保留一个）
    l.unique();
    //遍历链表
    for(it=l.begin();it!=l.end();it++)
    {
        cout<<*it<<" ";
    }
    //回车换行
    cout<<endl;
    return 0;
}
```

运行结果：

2 8 1 1 1 5 1
2 8 1 5 1

2.10 bitset 位集合容器

bitset 容器是一个 bit 位元素的序列容器，每个元素只占一个 bit 位，取值为 0 或 1，因而很节省内存空间。图 2-8 是一个 bitset 的存储示意图，它的 10 个元素只使用了两个字节的空间。

图 2-8 bitset 的存储示意图

使用 bitset，需要声明头文件包含语句"#include <bitset>"，bitset 文件在 C:\Program Files\Microsoft Visual Studio\VC98\Include 文件夹下。

bitset 类提供的方法见下表。

bitset 类的方法列表（bitset<n> b）

方法	功能
b.any()	b 中是否存在置为 1 的二进制位？
b.none()	b 中不存在置为 1 的二进制位吗？
b.count()	b 中置为 1 的二进制位的个数
b.size()	b 中二进制位的个数
b[pos]	访问 b 中在 pos 处的二进制位
b.test(pos)	b 中在 pos 处的二进制位是否为 1？
b.set()	把 b 中所有二进制位都置为 1
b.set(pos)	把 b 中在 pos 处的二进制位置为 1
b.reset()	把 b 中所有二进制位都置为 0
b.reset(pos)	把 b 中在 pos 处的二进制位置为 0
b.flip()	把 b 中所有二进制位逐位取反
b.flip(pos)	把 b 中在 pos 处的二进制位取反
b.to_ulong()	用 b 中同样的二进制位返回一个 unsigned long 值
os << b	把 b 中的位集输出到 os 流

2.10.1 创建 bitset 对象

创建 bitset 对象时，必须要指定容器的大小。bitset 对象的大小一经定义，就不能修改了。

下面这条语句就定义了 bitset 对象 b，它能容纳 100 000 个元素，即 100 000 个 bit（位），此时，所有元素的值都为 0。

```
bitset<100000> b
```

2.10.2 设置元素值

（1）采用下标法。

```
#include <bitset>
#include <iostream>
using namespace std;
int main(int argc, char * argv[])
{
    bitset<10> b;
    //采用下标法给元素赋值
    b[1]=1;
    b[6]=1;
    b[9]=1;
    //下标法输出所有元素，第 0 位是最低位，第 9 位是最高位
```

```
    int i;
    for(i=b.size()-1;i>=0;i--)
    {
        cout<<b[i];
    }
    cout<<endl;
    return 0;
}
```

运行结果：

1001000010

（2）采用 set()方法，一次性将元素设置为 1。

```
#include <bitset>
#include <iostream>
using namespace std;

int main(int argc, char * argv[])
{
    bitset<10> b;
    //采用 set()方法，一次性将元素设置为 1
    b.set();
    //下标法输出所有元素，第 0 位是最低位，第 9 位是最高位
    int i;
    for(i=b.size()-1;i>=0;i--)
    {
        cout<<b[i];
    }
    cout<<endl;
    return 0;
}
```

运行结果：

1111111111

（3）采用 set(pos)方法，将某 pos 位设置为 1。

```
#include <bitset>
#include <iostream>
using namespace std;

int main(int argc, char * argv[])
{
    bitset<10> b;
    //采用 set(pos)方法，将元素设置为 1
    b.set(1,1);
    b.set(6,1);
    b.set(9,1);
    //下标法输出所有元素，第 0 位是最低位，第 9 位是最高位
    int i;
```

```
    for(i=b.size()-1;i>=0;i--)
    {
        cout<<b[i];
    }
    cout<<endl;
    return 0;
}
```

运行结果：

1001000010

（4）采用 reset(pos) 方法，将某 pos 位设置为 0。

```
#include <bitset>
#include <iostream>
using namespace std;

int main(int argc, char * argv[])
{
    bitset<10> b;
    //采用set()方法，将元素全部设置为1
    b.set();
    //采用reset(pos)方法，将元素设置为0
    b.reset(0);
    b.reset(2);
    b.reset(3);
    b.reset(4);
    b.reset(5);
    b.reset(7);
    b.reset(8);
    //下标法输出所有元素，第0位是最低位，第9位是最高位
    int i;
    for(i=b.size()-1;i>=0;i--)
    {
        cout<<b[i];
    }
    cout<<endl;
    return 0;
}
```

运行结果：

1001000010

2.10.3 输出元素

（1）采用下标法输出元素。

```
#include <bitset>
#include <iostream>
using namespace std;
```

```
int main(int argc, char * argv[])
{
    bitset<10> b;
    //采用set()方法，将元素全部设置为1
    b.set();
    //采用set(pos)方法，将元素设置为0
    b.set(0,0);
    b.set(2,0);
    b.set(3,0);
    b.set(4,0);
    b.set(5,0);
    b.set(7,0);
    b.set(8,0);
    //下标法输出所有元素，第0位是最低位，第9位是最高位
    int i;
    for(i=b.size()-1;i>=0;i--)
    {
        cout<<b[i];
    }
    cout<<endl;
    return 0;
}
```

运行结果：

1001000010

（2）直接向输出流输出全部元素。

```
#include <bitset>
#include <iostream>
using namespace std;

int main(int argc, char * argv[])
{
    bitset<10> b;
    //采用set()方法，将元素全部设置为1
    b.set();
    //采用set(pos)方法，将元素设置为0
    b.set(0,0);
    b.set(2,0);
    b.set(3,0);
    b.set(4,0);
    b.set(5,0);
    b.set(7,0);
    b.set(8,0);
    //直接向输出流输出全部元素
    cout<<b<<endl;
    return 0;
}
```

运行结果:

1001000010

2.11 stack 堆栈容器

stack 堆栈是一个后进先出（Last In First Out，LIFO）的线性表，插入和删除元素都只能在表的一端进行。插入元素的一端称为栈顶（Stack Top），而另一端则称为栈底（Stack Bottom）。插入元素叫入栈（Push），元素的删除则称为出栈（Pop）。图 2-9 是堆栈示意图。

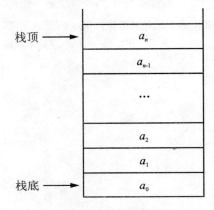

图 2-9　堆栈示意图

要使用 stack，必须声明头文件包含语句"#include <stack>"。stack 文件在 C:\Program Files\Microsoft Visual Studio\VC98\Include 文件夹中。

堆栈的使用方法

堆栈只提供入栈、出栈、栈顶元素访问和判断是否为空等几种方法。

采用 push() 方法将元素入栈；采用 pop() 方法出栈；采用 top() 方法访问栈顶元素；采用 empty() 方法判断堆栈是否为空，如果是空的，则返回逻辑真，否则返回逻辑假。当然，可以采用 size() 方法返回当前堆栈中有几个元素。

下面的程序是对堆栈各种方法的示例：

```
#include <stack>
#include <iostream>
using namespace std;

int main(int argc, char * argv[])
{
    //定义堆栈 s，其元素类型是整型
    stack<int> s;
    //元素入栈
    s.push(1);
    s.push(2);
    s.push(3);
```

```
    s.push(9);
    //读取栈顶元素
    cout<<s.top()<<endl;
    //返回堆栈元素数量
    cout<<s.size()<<endl;
    //判断堆栈是否为空
    cout<<s.empty()<<endl;
    //所有元素出栈（删除所有元素）
    while(s.empty()!=true)//堆栈非空
    {
        cout<<s.top()<<" ";//读取栈顶元素
        s.pop();//出栈（即删除栈顶元素）
    }
    //回车换行
    cout<<endl;
    return 0;
}
```

运行结果：

9
4
0
9 3 2 1

2.12 queue 队列容器

queue 队列容器是一个先进先出（First In First Out，FIFO）的线性存储表，元素的插入只能在队尾，元素的删除只能在队首。图 2-10 是 queue 队列容器数据结构示意图。

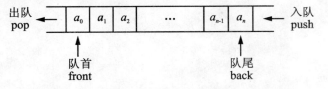

图 2-10 queue 队列容器数据结构示意图

使用 queue 需要声明头文件包含语句"#include <queue>"，queue 文件在 C:\Program Files\Microsoft Visual Studio\VC98\Include 文件夹里。

queue 队列的使用方法

queue 队列具有入队 push()（即插入元素）、出队 pop()（即删除元素）、读取队首元素 front()、读取队尾元素 back()、判断队列是否为空 empty()和队列当前元素的数目 size() 这几种方法。

下面的程序详细说明了 queue 队列的使用方法：

```
#include <queue>
#include <iostream>
using namespace std;
int main(int argc, char * argv[])
{
    //定义队列，元素类型是整型
    queue<int> q;
    //入队，即插入元素
    q.push(1);
    q.push(2);
    q.push(3);
    q.push(9);
    //返回队列元素数量
    cout<<q.size()<<endl;
    //队列是否为空，是空，则返回逻辑真，否则返回逻辑假
    cout<<q.empty()<<endl;
    //读取队首元素
    cout<<q.front()<<endl;
    //读取队尾元素
    cout<<q.back()<<endl;
    //所有的元素出列（删除所有元素）
    while(q.empty()!=true)
    {
        cout<<q.front()<<" ";
        //队首元素出队（删除队首元素）
        q.pop();
    }
    //回车换行
    cout<<endl;
    return 0;
}
```

运行结果：

4
0
1
9
1 2 3 9

2.13　priority_queue 优先队列容器

　　priority_queue 优先队列容器与队列一样，只能从队尾插入元素，从队首删除元素。但它有一个特性，就是队列中最大的元素总是位于队首，所以出队时，并非按先进先出的原则进行，而是将当前队列中最大的元素出队。这点类似于给队列里的元素进行了由大到小

的顺序排序。元素的比较规则默认为按元素的值由大到小排序；当然，可以重载"<"操作符来重新定义比较规则。

使用 priority_queue 需要声明头文件包含语句"#include <queue>"，它与 queue 队列共用一个头文件，queue 文件在 C:\Program Files\Microsoft Visual Studio\VC98\Include 文件夹中。

2.13.1 优先队列的使用方法

优先队列包含入队 push()（插入元素）、出队 pop()（删除元素）、读取队首元素 top()、判断队列是否为空 empty() 和读取队列元素数量 size() 等方法。

下面这个程序详细说明了优先队列的使用方法：

```cpp
#include <queue>
#include <iostream>
using namespace std;
int main(int argc, char * argv[])
{
    //定优先队列，元素类型为整型
    priority_queue<int> pq;
    //入队，插入新元素
    pq.push(1);
    pq.push(2);
    pq.push(3);
    pq.push(9);
    //返回队列中元素数目
    cout<<pq.size()<<endl;
    //所有元素出队，删除所有元素
    while(pq.empty()!=true)
    {
        //读取当前队首元素
        cout<<pq.top()<<" ";
        //出队，删除队首元素
        pq.pop();
    }
    //回车换行
    cout<<endl;
    return 0;
}
```

运行结果：

4
9 3 2 1

2.13.2 重载"<"操作符来定义优先级

如果优先队列的元素类型是结构体，可以通过在结构体中重载"<"操作符的方法来修改优先队列的优先性。

下面这个程序很好地说明了如何在结构体中重载"<"操作符:

```cpp
#include <queue>
#include <string>
#include <iostream>
using namespace std;

//定义结构体
struct Info{
    string name;
    float score;
    //重载"<"操作符,指定优先规则(排序规则)
    bool operator < (const Info &a) const
    {
        //按 score 由小到大排列。如果要由大到小排列,使用">"号即可
        return a.score<score;
    }
};
int main(int argc, char * argv[])
{
    //定义优先队列,元素类型为 Info 结构体
    priority_queue<Info> pq;
    //定义结构体变量
    Info info;

    //入队
    info.name="Jack";
    info.score=68.5;
    pq.push(info);

    info.name="Bomi";
    info.score=18.5;
    pq.push(info);

    info.name="Peti";
    info.score=90;
    pq.push(info);
    //元素全部出队
    while(pq.empty()!=true)
    {
        //返回队首元素
        cout<<pq.top().name<<" : "<<pq.top().score<<endl;
        //出队,删除队首元素
        pq.pop();
    }
    return 0;
}
```

运行结果:

```
Bomi : 18.5
Jack : 68.5
Peti : 90
```

2.13.3 重载"()"操作符来定义优先级

如果优先队列的元素不是结构体类型，则可以通过重载"()"操作符的方式来定义优先级。当然，元素是结构体类型，也可以通过重载"()"操作符的方式来定义优先级，而不是一定要在结构体内重载"<"操作符才行。

下面这个程序说明了如何重载"()"操作符：

```cpp
#include <queue>
#include <vector>
#include <iostream>
using namespace std;
//重载"()"操作符
struct myComp
{
    bool operator()(const int &a,const int &b)
    {
        //由小到大排列采用">"号；如果要由大到小排列，则采用"<"号
        return a>b;
    }
};
int main(int argc, char * argv[])
{
    //定义优先队列，元素类型为Info结构体，显式说明内部结构是vector
    priority_queue<int,vector<int>,myComp> pq;
    //入队
    pq.push(1);
    pq.push(9);
    pq.push(2);
    pq.push(30);
    //元素全部出队
    while(pq.empty()!=true)
    {
        //返回队首元素
        cout<<pq.top()<<" ";
        //出队，删除队首元素
        pq.pop();
    }
    cout<<endl;
    return 0;
}
```

运行结果：

1 2 9 30

第 3 章　ACM 程序设计基础

3.1　读入一个参数

3.1.1　链接地址

http://www.realoj.com/网上第 68 题

3.1.2　题目内容

已知正方形的边长，试编程求出其面积。

输入描述：输入不超过 50 个正整数的数据 n（$1 \leqslant n \leqslant 10\,000$），每个正整数间以空格隔开。

输出描述：每次读入一个正整数，便输出其正方形的面积数，输出每个面积后再回车。

输入样例

　1　3　5　7

输出样例

　1
　9
　25
　49

3.1.3　参考答案

本章所有程序中的"ifstream cin("aaa.txt");"语句，是作为本机调试使用的，在网上提交时，一定要用"//"注释掉。后面的程序也都这样做。

```
#include <fstream>
#include <iostream>
using namespace std;

int main(int argc, char * argv[])
{
   ifstream cin("aaa.txt");
   int n;
   while(cin>>n)
   {
      cout<<n*n<<endl;
   }
   return 0;
}
```

3.2 读入两个参数

3.2.1 链接地址

http://www.realoj.com/网上第 69 题

3.2.2 题目内容

编写程序计算两个整数的差。

输入描述：输入数据含有不超过 50 个整数对，每个整数及每对整数的运算结果都不会超过 $\pm 2^{31}$。

输出描述：对于每次读入的一对整数，输出前者减去后者的差。每个结果以回车结束。

输入样例

```
1 3 5 7
```

输出样例

```
-2
-2
```

3.2.3 参考答案

```cpp
#include <fstream>
#include <iostream>
using namespace std;
int main(int argc, char * argv[])
{
    ifstream cin("aaa.txt");
    int a,b;
    while(cin>>a>>b)
    {
        cout<<a-b<<endl;
    }
    return 0;
}
```

3.3 1!到 n!的和

3.3.1 链接地址

http://www.realoj.com/网上第 70 题

3.3.2 题目内容

求 1!+2!+3!+4!+⋯+n!的结果。

输入描述：输入不多于 50 个正整数的数据 n（1≤n≤12）。

输出描述：对于每个 n，输出计算结果。每个计算结果应单独占一行。

输入样例

```
3 6
```

输出样例

```
9
873
```

3.3.3 参考答案

```cpp
#include <fstream>
#include <iostream>
using namespace std;
int main(int argc, char * argv[])
{
    ifstream cin("aaa.txt");
    int n,sum,p;
    while(cin>>n)
    {
        sum=0;
        p=1;
        for(int i=1;i<=n;i++)
        {
            p=p*i;
            sum=sum+p;
        }
        cout<<sum<<endl;
    }
    return 0;
}
```

3.4 等比数列

3.4.1 链接地址

http://www.realoj.com/网上第 71 题

3.4.2 题目内容

已知 q 与 n，求等比数列之和：$1+q+q^2+q^3+q^4+\cdots+q^n$。

输入描述：输入数据不多于 50 对，每对数据含有一个整数 n（$1\leq n\leq 20$）、一个小数 q（$0<q<2$）。

输出描述：对于每组数据 n 和 q，计算其等比数列的和，精确到小数点后 3 位，每个计算结果应单独占一行。

输入样例

```
6 0.3 5 1.3
```

输出样例

```
1.428
12.756
```

3.4.3 参考答案

对于等比数列之和 $S_n = a_1 + a_2 + a_3 + \cdots + a_n$，有公式 $S_n = \dfrac{a_1(1-q^n)}{1-q}$（$q \neq 1$）。本例要求的等比数列，实际上是有 $n+1$ 项，且 $a_1 = 1$。

另外，求 x^y 的函数是 pow(x,y)，它需要 "#include <cmath>"。

本题要求控制小数点后的位数，如果采用 C 语言的 printf 函数来输出，那么控制小数点后的位数形式为 "printf("%*.*f", a);"，如 "printf("%.3f\n",sum);"。如果采用 C++的 cout 输出，那么先用 "cout.precision(n);" 来设定小数点后保留 n 位，然后，输出时加 "fixed" 参数，表明是定点输出。

```cpp
#include <iostream>
#include <fstream>
#include <cmath>
using namespace std;
int main(int argc, char * argv[])
{
    ifstream cin("aaa.txt");
    //定点输出小数点后 3 位；否则是输出有效数字 3 位
    cout.precision(3);
    int n;
    double q,sum;
    while(cin>>n>>q)
    {
        if(q==1)sum=1+n;
        else
            sum=(1-pow(q,n+1))/(1-q);
        //使用 fixed，定点输出，即小数点后的精度是 3 位
        cout<<fixed<<sum<<endl;
    }
    return 0;
}
```

3.5 菲波那契数

3.5.1 链接地址

http://www.realoj.com/网上第 72 题

3.5.2 题目内容

菲波那契（Fibonacci）数（简称菲氏数）定义为：

$$\begin{cases} f(0) = 0 \\ f(1) = 1 \\ f(n) = f(n-1) + f(n-2) \quad (n > 1 \text{且} n \in \text{整数}) \end{cases}$$

如果写出菲氏数列，则应该是：

  0 1 1 2 3 5 8 13 21 34 …

如果求其第 6 项，则应为 8。

求第 n 项菲氏数。

输入描述：输入数据含有不多于 50 个的正整数 n（$0 \leq n \leq 46$）。

输出描述：对于每个 n，计算其第 n 项菲氏数，每个结果应单独占一行。

输入样例

  6 10

输出样例

  8
  55

3.5.3 参考答案

先把第 0 项到第 46 项的菲波那契数求出来，放在一个表（或数组）中，然后，直接查表即可，这样就不会超时。

（1）采用数组来做。

```cpp
#include <iostream>
#include <fstream>
#include <cmath>

using namespace std;
int main(int argc, char * argv[])
{
    ifstream cin("aaa.txt");
    int a[47];
    a[0]=0;
```

```
    a[1]=1;
//先把前 46 项菲波那契数求出来放在表（或数组）中
    for(int i=2;i<=46;i++)
    {
        a[i]=a[i-1]+a[i-2];
    }
//查表（数组）
    int n;
    while(cin>>n)
    {
        cout<<a[n]<<endl;
    }
    return 0;
}
```

（2）采用向量来做。

```
#include <fstream>
#include <iostream>
#include <vector>

using namespace std;

int main(int argc, char * argv[])
{
    ifstream cin("aaa.txt");
    vector<unsigned int>v;
    unsigned int n;
//先建表，把第 0~46 项的菲波那契数都算出来
    v.push_back(0);
    v.push_back(1);
    for(int i=2;i<=46;i++)
    {
        v.push_back(v[i-1]+v[i-2]);
    }
//直接输出第 n 项菲波那契数（从 0 项开始计算）
    while(cin>>n)
    {
        cout<<v[n]<<endl;
    }
    return 0;
}
```

3.6 最大公约数

3.6.1 链接地址

http://www.realoj.com/网上第 73 题

3.6.2 题目内容

求两个正整数的最大公约数。

输入描述：输入数据含有不多于 50 对的数据，每对数据由两个正整数（$0 < n_1, n_2 < 2^{32}$）组成。

输出描述：对于每组数据 n_1 和 n_2，计算最大公约数，每个计算结果应占单独一行。

输入样例

```
6 5 18 12
```

输出样例

```
1
6
```

3.6.3 参考答案

求两数的最大公约数，可采用欧几里得方法：只要两数不相等，就反复用大数减小数，直到相等为止，此相等的数就是两数的最大公约数。

```cpp
#include <iostream>
#include <fstream>

using namespace std;
//声明 gcd 函数，该函数用来计算两数的最大公约数
int gcd(int,int);
int main(int argc, char * argv[])
{
    ifstream cin("aaa.txt");
    int x,y;
    while(cin>>x>>y)
    {
        cout<<gcd(x,y)<<endl;
    }
    return 0;
}
int gcd(int x,int y)
{
    while(x!=y)
    {
        if(x>y)x=x-y;
        else
            y=y-x;
    }
    return x;
}
```

3.7 最小公倍数

3.7.1 链接地址

http://www.realoj.com/网上第 74 题

3.7.2 题目内容

求两个正整数的最小公倍数。

输入描述：输入数据含有不多于 50 对的数据，每对数据由两个正整数（$0<n_1$，$n_2<100\,000$）组成。

输出描述：对于每组数据 n_1 和 n_2，计算最小公倍数，每个计算结果应单独占一行。

输入样例

```
6 5 18 12
```

输出样例

```
30
36
```

3.7.3 参考答案

对于 32 位 CPU，int 的表示范围为 $-2^{31} \sim 0 \sim (+2^{31}-1)$，即 $-2\,147\,483\,648 \sim 2\,147\,483\,647$ 有符号数，用最高位表示符号（1 正 0 负）。

最小公倍数=X*Y/gcd(x,y)；但两数先乘会产生很大的数，可能会超过整数的表示范围，所以，把计算顺序修改一下就可以了：最小公倍数=X/gcd(x,y)*Y。

```cpp
#include <iostream>
#include <fstream>

using namespace std;
//声明 gcd 函数，该函数用来计算两数的最大公约数
int gcd(int,int);
int main(int argc, char * argv[])
{
    ifstream cin("aaa.txt");
    int x,y;
    while(cin>>x>>y)
    {
        cout<<x/gcd(x,y)*y<<endl;
    }
    return 0;
}
int gcd(int x,int y)
{
```

```
    while(x!=y)
    {
        if(x>y)x=x-y;
        else
            y=y-x;
    }
    return x;
}
```

3.8 平均数

3.8.1 链接地址

http://www.realoj.com/网上第 75 题

3.8.2 题目内容

求若干个整数的平均数。

输入描述：输入数据含有不多于 5 组的数据，每组数据由一个整数 n（$n \leq 50$）打头，表示后面跟着 n 个整数。

输出描述：对于每组数据，输出其平均数，精确到小数点后 3 位，每个平均数应单独占一行。

输入样例

```
3 6 5 18
4 1 2 3 4
```

输出样例

```
9.667
2.500
```

3.8.3 参考答案

求平均数，保留小数后三位，需要四舍五入，方法如下：

```
cout<<precision(3);
cout<<fixed<<sum/n<<endl;

#include <iostream>
#include <fstream>
using namespace std;
int main(int argc, char * argv[])
{
    ifstream cin("aaa.txt");
    int n,x;
    double sum;
```

```
cout.precision(3);//输出精确到小数后三位,四舍五入
while(cin>>n)
{
    sum=0;
    for(int i=1;i<=n;i++)
    {
        cin>>x;
        sum=sum+x;
    }
    //定点输出小数点后三位
    cout<<fixed<<sum/n<<endl;
}
return 0;
}
```

3.9 对称三位数素数

3.9.1 链接地址

http://www.realoj.com/网上第 76 题

3.9.2 题目内容

判断一个数是否为对称三位数素数。所谓"对称"是指一个数,倒过来还是该数。例如,375 不是对称数,因为倒过来变成了 573。

输入描述:输入数据含有不多于 50 个的正整数（$0<n<2^{32}$）。

输出描述:对于每个 n,如果该数是对称三位数素数,则输出"Yes",否则输出"No"。每个判断结果单独列一行。

输入样例

```
11 101 272
```

输出样例

```
No
Yes
No
```

3.9.3 参考答案

素数是指只能被 1 和本身整除的自然数（1 不是素数）,前几位素数是 2, 3, 5, 7, 11, 13, 17, 19, 23, 27, 29, 31。

```
#include <iostream>
#include <fstream>
#include <cmath>
```

```
using namespace std;
//isPrime 函数用来判断一个数是否是素数
bool isPrime(int);
int main(int argc, char * argv[])
{
    ifstream cin("aaa.txt");
    int n;
    while(cin>>n)
    {
        cout<<(n>100&&n<1000&&n/100==n%10&&isPrime(n)?"Yes\n":"No\n");
    }
    return 0;
}
bool isPrime(int n)
{
    //sqrt 开方，需要#include <cmath>
    int sqr=sqrt(n*1.0);
    for(int i=2;i<=sqr;i++)
    {
        if(n%i==0)return false;
    }
    return true;
}
```

3.10 十进制转换为二进制

3.10.1 链接地址

http://www.realoj.com/网上第 77 题

3.10.2 题目内容

将十进制整数转换成二进制数。

输入描述：输入数据中含有不多于 50 个的整数 n（$-2^{31}<n<2^{31}$）。

输出描述：对于每个 n，以 11 位的宽度右对齐输出 n 值，然后输出"-->"，再然后输出二进制数。每个整数 n 的输出，独立占一行。

输入样例

```
2
0
-12
1
```

输出样例

```
          2-->10
          0-->0
```

```
            -12-->-1100
              1-->1
```

3.10.3 参考答案

```
#include <iostream>
#include <fstream>
#include <string>
#include <algorithm>
using namespace std;

string s;//全局变量
int main(int argc, char * argv[])
{
   ifstream cin("aaa.txt");
    int n;
   while(cin>>n)
   {
      if(n==0){
          cout<<"          0-->0\n";
          continue;
      }
      s="";
      for(int a=n;a;a=a/2){
          s=s+(a%2?'1':'0');
      }
      std::reverse(s.begin(),s.end());
      const char *sss=s.c_str();
      cout.width(11);//输出宽度为11
      cout<<n<<(n<0?"-->-":"-->")<<sss<<"\n";
   }
    return 0;
}
```

3.11 列出完数

3.11.1 链接地址

http://www.realoj.com/网上第 78 题

3.11.2 题目内容

自然数中，完数寥若晨星，请在从 1 到某个整数范围中打印出所有的完数来。所谓"完数"是指一个数恰好等于它所有不同因子之和。例如，6 是完数，因为 6=1+2+3。而 24 不是完数，因为 24≠1+2+3+4+6+8+12（=36）。

输入描述：输入数据中含有一些整数 n（$1<n<10\,000$）。

输出描述：对于每个整数 n，输出所有不大于 n 的完数。每个整数 n 的输出由 n 引导，跟上冒号，然后是由空格开道的一个个完数，每个 n 的完数列表应占独立的一行。

输入样例

```
100
5000
```

输出样例

```
100: 6 28
5000: 6 28 496
```

3.11.3 参考答案

本题限时一秒。先把小于 10 000 的完数计算出来，放在向量中，然后对于每个 n，只需在向量中打印每个完数，直到 n 之内的完数的界即可。这种算法不会超时。

```cpp
#include <fstream>
#include <iostream>
#include <vector>
using namespace std;

int main(int argc, char * argv[])
{
    ifstream cin("aaa.txt");
    vector<int>a;
    for(int i=2;i<10000;i=i+2)
    {
        int sum=1;
        for(int j=2;j<=i/2;j++)
        {
            if(i%j==0)sum=sum+j;
        }
        if(sum==i)a.push_back(i);
    }

    int n;
    while(cin>>n)
    {
        cout<<n<<":";
        for(int i=0;i<a.size();i++)//推测完数为偶数，故步长为2
        {
            if(a[i]<=n)cout<<" "<<a[i];
        }
        cout<<endl;
    }
    return 0;
}
```

3.12 12！配　对

3.12.1 链接地址

http://www.realoj.com/网上第 79 题

3.12.2 题目内容

找出输入数据中所有两两相乘的积为 12!的个数。

输入描述：输入数据中含有一些整数 n（$1 \leqslant n < 2^{32}$）。
输出描述：输出所有两两相乘的积为 12!的个数。
输入样例

```
1 10000 159667200 9696 38373635
1000000 479001600 3
```

输出样例

```
2
```

3.12.3 参考答案

```cpp
#include <fstream>
#include <iostream>
#include <set>
using namespace std;
int main(int argc, char * argv[])
{
    ifstream cin("aaa.txt");
    int num=0;
    int f12=479001600;
    //多重集合，允许值重复
    multiset<unsigned int>s;
    int n;
    while(cin>>n)
    {
        if(f12%n==0)//n 是 f12 的约数吗
        {
            //多重集合中有 n 的因子吗
            multiset<unsigned int>::iterator it=s.find(f12/n);
            if(it!=s.end())
            {
                num++;
                s.erase(it);//从多重集合中删除该因子
            }
```

```
        else
            s.insert(n);//插入到多重集合中
        }
    }
    cout<<num<<endl;//输出因式对数
    return 0;
}
```

3.13 五位以内的对称素数

3.13.1 链接地址

http://www.realoj.com/网上第 80 题

3.13.2 题目内容

判断一个数是否为对称且不大于五位数的素数。

输入描述：输入数据含有不多于 50 个的正整数 n（$0<n<2^{32}$）。

输出描述：对于每个 n，如果该数是不大于五位数的对称素数，则输出"Yes"，否则输出"No"。每个判断结果单独列一行。

输入样例

```
11 101 272
```

输出样例

```
Yes
Yes
No
```

3.13.3 参考答案

```
#include <fstream>
#include <iostream>

using namespace std;

bool isPrime(int n)//判断是否是素数
{
    if(n==1)return false;//1 不是素数
    if(n!=2&&n%2==0)return false;//2 是素数,清除 2 的倍数
    for(int i=3;i*i<=n;i=i+2)//2,3,5,7 是素数
    {
        if(n%i==0)return false;
    }
    return true;
}
bool isSym(int n)
```

```
{
    //一位数的素数和 11 都算是对称的
    if(n<12&&n!=10)return true;
    //三位数素数是否对称
    if(n>100 && n<1000 && n/100==n%10)return true;
    //四位数的对称数,不可能是素数
    //假设四位对称数 abba,即 1000a+100b+10b+a=1001a+110b=11(91a+10b)
    //具有因子 11,是合数
    //判断五位素数是否对称
    if(n>10000 && n/1000==n%10*10+n/10%10)return true;
    return false;
}
int main(int argc, char * argv[])
{
    ifstream cin("aaa.txt");
    int n;
    while(cin>>n)
    {
        cout<<(n<100000 && issym(n)  && isPrime(n)? "Yes\n":"No\n");
    }
    return 0;
}
```

3.14 01 串 排 序

3.14.1 链接地址

http://www.realoj.com/网上第 81 题

3.14.2 题目内容

将 01 串首先按长度排序,长度相同时,按 1 的个数多少进行排序,1 的个数相同时再按 ASCII 码值排序。

输入描述:输入数据中含有一些 01 串,01 串的长度不大于 256 个字符。
输出描述:重新排列 01 串的顺序,使得串按题目描述的方式排序。

输入样例

10011111
00001101
1010101
1
0
1100

输出样例

0
1

1100
1010101
00001101
10011111

3.14.3 参考答案

```cpp
#include <fstream>
#include <iostream>
#include <string>
#include <set>
#include <algorithm>

using namespace std;

struct Comp
{
    bool operator()(const string &s1,const string &s2)
    {
        if(s1.length()!=s2.length())return s1.length()<s2.length();
        int c1=count(s1.begin(),s1.end(),'1');
        int c2=count(s2.begin(),s2.end(),'1');
        return (c1!=c2?c1<c2:s1<s2);
    }
};

int main(int argc, char * argv[])
{
    multiset<string,Comp>ms;
    ifstream cin("aaa.txt");
    string s;
    while(cin>>s)
    {
        ms.insert(s);
    }
    for(multiset<string,Comp>::iterator it=ms.begin();it!=ms.end();it++)
    {
        cout<<*it<<endl;
    }
    return 0;
}
```

3.15 排列对称串

3.15.1 链接地址

http://www.realoj.com/ 网上第 82 题

3.15.2 题目内容

字符串有些是对称的，有些是不对称的，请将那些对称的字符串按从小到大的顺序输出。字符串先以长度论大小，如果长度相同，再以 ASCII 码值为排序标准。

输入描述：输入数据中含有一些字符串（1≤串长≤256）。

输出描述：根据每个字符串，输出对称的那些串，并且要求按从小到大的顺序输出。

输入样例

```
123321
123454321
123
321
sdfsdfd
121212
@@dd@@
```

输出样例

```
123321
@@dd@@
123454321
```

3.15.3 参考答案

```cpp
#include <fstream>
#include <iostream>
#include <string>
#include <vector>
#include <algorithm>

using namespace std;

//自己设计排序比较函数
bool Comp(const string &s1,const string &s2)
{
    return s1.length()!=s2.length()?s1.length()<s2.length():s1<s2;
}
int main(int argc, char * argv[])
{
    ifstream cin("aaa.txt");
    vector<string>v;
    string t,s;
    while(cin>>s)
    {
        t=s;
        //反转字符串，用来判断字符是否对称
        reverse(t.begin(),t.end());
        if(t==s)
        {
            v.push_back(s);
```

```
        }
    }
    //按Comp函数比较规则排序
    sort(v.begin(),v.end(),Comp);
    //输出向量中的所有元素
    for(int i=0;i<v.size();i++)
    {
        cout<<v[i]<<endl;
    }
    return 0;
}
```

3.16 按绩点排名

3.16.1 链接地址

http://www.realoj.com/网上第83题

3.16.2 题目内容

有一些班级的学生需要按绩点计算并排名。每门课程的成绩只有在60分以上（含），才予以计算绩点。课程绩点的计算公式为：(课程成绩–50)÷10×学分数。一个学生的总绩点为其所有课程绩点总和除以10。

输入描述：输入数据中含有一些班级（数量≤20）。每个班级的第一行数据 n（≤10），$a_1, a_2, a_3, \cdots, a_n$，表示该班级共有 n 门课程，每门课程的学分分别为 $a_1, a_2, a_3, \cdots, a_n$；班级数据中的第二行数据为一个整数 m（≤50），表示本班级有 m 个学生；班级数据接下去有 m 行对应 m 个学生数据；每行学生数据中的第一个为字串 s_1（s_1 中间没有空格），表示学生姓名，后面跟有 n 个整数 $s_1, s_2, s_3, \cdots, s_n$，表示该学生各门课程的成绩（$0 \leq s_i \leq 100$）。

输出描述：以班级为单位输出各个学生按绩点分从大到小的排名。如果绩点分相同，则按学生名字的ASCII串值从小到大排名。每个班级的排名输出之前应先给出一行，描述班级序号"class #:"（#表示班级序号），班级之间应空出一行。排名时，每个学生占一行，列出名字和总绩点。学生输出宽度为10个字符，左对齐，在空出一格后列出总绩点。

输入样例

```
1
3 3 4 3
3
张三    89 62 71
Smith   98 50 80
王五    67 88 91
```

输出样例

```
class 1:
王五        3.26
```

```
Smith      2.34
张三       2.28
```

3.16.3 参考答案

```cpp
#include <fstream>
#include <iostream>
#include <string>
#include <vector>
#include <algorithm>
#include <iomanip>//cout 的输出格式

using namespace std;

struct student
{
    string s;
    double d;
};
//编写比较函数
bool myComp(const student &s1,const student &s2)
{
        if(s1.d!=s2.d)return s1.d>s2.d;//由大到小则使用">"号
        if(s1.s!=s2.s)return s1.s<s2.s;//由小到大则使用"<"号
}

int main(int argc, char * argv[])
{
    ifstream cin("aaa.txt");
    int n;//班级数量
    int c;//课程数量
    double xf;//学分
    vector<double>vxf;//学分向量
    int p;//班级人数
    string name;//学生名称
    double score;//成绩
    student xs;//学生名称与总学分结构体
    vector<student>vxs;//最终学生名称与总学分
    cin>>n;//n 个班
    for(int i=0;i<n;i++)//处理每一个班
    {
        cin>>c;//读入课程数量
        for(int j=0;j<c;j++)//读入每门课程的学分
        {
            cin>>xf;
            vxf.push_back(xf);
        }
        cin>>p;//读入班级人数
        //读入一个班的 p 个学生的名称与每门课成绩
        for(int k=0;k<p;k++)
```

```
        {
            cin>>name;//读入学生名称
            xs.s=name;
            xf=0.0;
            for(int m=0;m<c;m++)//读入每门课程的分数
            {
                cin>>score;
                //成绩要大于或等于60分才计算绩点
                if(score<60)continue;
                xf=xf+(score-50)/10*vxf[m];//计算总学分
            }
            xs.d=xf/10;
            vxs.push_back(xs);
        }
        //输出每个班的情况
        cout<<(i?"\n":"");
        cout<<"class "<<i+1<<":"<<endl;
        sort(vxs.begin(),vxs.end(),myComp);//带自定义比较函数myComp
        for(vector<student>::iterator it=vxs.begin();it<vxs.end();it++)
        {
            cout<<fixed<<setprecision(2);
            cout<<left<<setw(11);
            cout<<(*it).s<<(*it).d<<endl;
        }
        vxf.clear();//清除向量
        vxs.clear();//清除向量
    }
    return 0;
}
```

3.17 按 1 的个数排序

3.17.1 链接地址

http://www.realoj.com/网上第 84 题

3.17.2 题目内容

有一些 0、1 字符串，将其按 1 的个数的多少的顺序进行输出。

输入描述：本题只有一组测试数据。输入数据由若干行组成，每行是一个数字，它是由若干个 0 和 1 组成的数字。

输出描述：对所有输入的数据，按 1 的个数进行升序排序，每行输出一个数字。
输入样例

```
10011111
00001101
```

```
1010101
1
0
1100
```

输出样例

```
0
1
1100
00001101
1010101
10011111
```

3.17.3 参考答案

```cpp
#include <fstream>
#include <iostream>
#include <string>
#include <vector>
#include <algorithm>

using namespace std;
bool myComp(const string &s1,const string &s2)
{
    int c1=count(s1.begin(),s1.end(),'1');
    int c2=count(s2.begin(),s2.end(),'1');
    //彻底修改排序规则，只按1的数量排序，
    //如果1的数量相等，则按出现的先后顺序
    //否则，会按ASCII码大小排序
    //只能用">"或"<"，不能用"="
    return c1!=c2?c1<c2:c1<c2;
}

int main(int argc, char * argv[])
{
    ifstream cin("aaa.txt");
    vector<string>vstr;
    string str;
    while(cin>>str)
    {
        vstr.push_back(str);
    }
    sort(vstr.begin(),vstr.end(),myComp);
    for(vector<string>::iterator it=vstr.begin();it<vstr.end();it++)
        cout<<*it<<endl;
    return 0;
}
```

第 4 章 ACM 程序设计实战

4.1 Quicksum

4.1.1 链接地址

http://www.realoj.com/网上第 85 题

4.1.2 时空限制

Time Limit: 1000 ms   Resident Memory Limit: 1024 KB   Output Limit: 1024 B

4.1.3 题目内容

A checksum is an algorithm that scans a packet of data and returns a single number. The idea is that if the packet is changed, the checksum will also change, so checksums are often used for detecting transmission errors, validating document contents, and in many other situations where it is necessary to detect undesirable changes in data.

For this problem, you will implement a checksum algorithm called Quicksum. A Quicksum packet allows only uppercase letters and spaces. It always begins and ends with an uppercase letter. Otherwise, spaces and letters can occur in any combination, including consecutive spaces.

A Quicksum is the sum of the products of each character's position in the packet times the character's value. A space has a value of zero, while letters have a value equal to their position in the alphabet. So, A=1, B=2, etc., through Z=26. Here are example Quicksum calculations for the packets "ACM" and "MID CENTRAL":

ACM: $1*1 + 2*3 + 3*13 = 46$

MID CENTRAL: $1*13 + 2*9 + 3*4 + 4*0 + 5*3 + 6*5 + 7*14 + 8*20 + 9*18 + 10*1 + 11*12 = 650$

Input

The input consists of one or more packets followed by a line containing only # that signals the end of the input. Each packet is on a line by itself, does not begin or end with a space, and contains from 1 to 255 characters.

Output

For each packet, output its Quicksum on a separate line in the output.

Example Input	Example Output
ACM	46
MID CENTRAL	650
REGIONAL PROGRAMMING CONTEST	4690
ACN	49
A C M	75
ABC	14
BBC	15
#	

4.1.4 题目来源

Mid-Central USA 2006

4.1.5 解题思路

本题要求计算一个数据包（即一行字符串）的 Quicksum。所谓 Quicksum，就是指一行字符串（数据包）中每个字符的位置与该字符的值的乘积相加的结果。一个数据包占一行，仅由大写字母和空格组成；位置由 1 开始计数，空格也占一个位置；A～Z 的值对应为 1～26，空格的值为 0。

本题的难点就是数据的读入。可采用 cin.getline()一行一行读入数据；也可以采用 cin.get()一个一个读入字符，但需要注意的是，cin.get()不会忽略任何字符，对于回车符需要单独处理。

4.1.6 参考答案

本章所有程序中的"ifstream cin("aaa.txt");"文件读入语句，是作为本机调试使用的，在网上提交时，一定要注释掉。后面的题目，也都是要这样做。

（1）下面是 cin.get()版本，字符一个一个读入。

```cpp
#include <fstream>
#include <iostream>
using namespace std;
int main(int argc, char * argv[])
{
    ifstream cin("aaa.txt");
    char ch;
    int i=1;
    int sum=0;
    //cin 会忽略回车、空格、Tab 跳格
    //采用 cin.get()一个一个读，就不会忽略任何字符
    //也可以采用 cin.getline()一行一行读入
    while(cin.get(ch))
    {
        if(ch=='#')break;//读完全部输入
```

```
            if(ch!='\n')//没读完一行
            {
                //空格字符不用计算
                if(ch!=' ')sum=sum+i*(ch-64);
                i++;
            }
            if(ch=='\n')
            {
                cout<<sum<<endl;
                sum=0;
                i=1;
            }
        }
        return 0;
    }
```

（2）下面是 cin.getline()版本，一次读入一行。

```
#include <fstream>
#include <iostream>
using namespace std;
int main(int argc, char * argv[])
{
    ifstream cin("aaa.txt");
    char ch[256];
    int i=1;
    int sum=0;
    //cin 会忽略回车、空格、Tab 跳格
    //采用 cin.get()一个一个读，就不会忽略任何字符
    //也可以采用 cin.getline()一行一行读入
    while(cin.getline(ch,256))
    {
        if(ch[0]=='#')break;
        for(int i=0;ch[i]!='\0';i++)
            if(ch[i]!=' ')sum=sum+(i+1)*(ch[i]-64);
        cout<<sum<<endl;
        sum=0;
    }
    return 0;
}
```

4.1.7 汉语翻译

1. 题目

Quicksum

checksum 是扫描一个数据包并返回一个数值的一种算法。其思路在于，如果数据包被修改过，那么，checksum 也会立即变化。所以，checksum 常常用于侦查数据传输错误，证实文档内容的完整性和其他需要检查数据不被修改的场合。

在本题中，你将实现一个 checksum 算法即 Quicksum。一个 Quicksum 数据包仅允许包含大写字母和空格，它通常是由一个大写字母开始和结束。然而，空格和字母可以出现在其他的位置中，连续的空格也是允许的。

Quicksum 是一行字符串（数据包）中每个字符的位置与该字符的值的乘积之和。空格的值是 0，字母的值等于它在字母表中的位置。所以，A 的值是 1，B 的值是 2，以此类推，Z 的值是 26。下面两个例子是求"ACM"和"MID CENTRAL"的 Quicksum：

ACM: 1*1 + 2*3 + 3*13 = 46

MID CENTRAL: 1*13 + 2*9 + 3*4 + 4*0 + 5*3 + 6*5 + 7*14 + 8*20 + 9*18 + 10*1 + 11*12 = 650

2. 输入描述

输入数据包含一个或多个数据包，输入数据以"#"结束。每个数据包占一行，不能以空格开始或结束，包含 1～255 个字符。

3. 输出描述

对于每个数据包，在每一行上输出它的 Quicksum。

输入样例	输出样例
ACM	46
MID CENTRAL	650
REGIONAL PROGRAMMING CONTEST	4690
ACN	49
A C M	75
ABC	14
BBC	15
#	

4.2　IBM Minus One

4.2.1　链接地址

http://www.realoj.com/网上第 86 题

4.2.2　时空限制

Time Limit: 1000 ms　Resident Memory Limit: 1024 KB　Output Limit: 1024 B

4.2.3　题目内容

You may have heard of the book *2001—A Space Odyssey* by Arthur C. Clarke, or the film of the same name by Stanley Kubrick. In it a spaceship is sent from Earth to Saturn. The crew is put into stasis for the long flight, only two men are awake, and the ship is controlled by the intelligent computer HAL. But during the flight HAL is acting more and more strangely, and even starts to kill the crew on board. We don't tell you how the story ends, in case you want to read the book for yourself :-)

After the movie was released and became very popular, there was some discussion as to what the name "HAL" actually meant. Some thought that it might be an abbreviation for "Heuristic Algorithm". But the most popular explanation is the following: if you replace every letter in the word HAL by its successor in the alphabet, you get ... IBM.

Perhaps there are even more acronyms related in this strange way! You are to write a program that may help to find this out.

Input

The input starts with the integer n on a line by itself—this is the number of strings to follow. The following n lines each contain one string of at most 50 upper-case letters.

Output

For each string in the input, first output the number of the string, as shown in the sample output. The print the string start is derived from the input string by replacing every time by the following letter in the alphabet, and replacing "Z" by "A".

Print a blank line after each test case.

Sample Input

```
2
HAL
SWERC
```

Sample Output

```
String #1
IBM

String #2
TXFSD
```

4.2.4 题目来源

Southwestern Europe 1997, Practice

4.2.5 解题思路

本题要求把一个字符串（由大写字母组成）转换为另一个字符串，转换的规则是，把字符串中的每个字符转换为字母表中的下一个字符，"Z"转换为"A"。

本题的难点在字符串的输出格式，即每行输出前要先输出"String #"行，每个测试案例后面要输出一个空行。输出格式在 ACM 程序设计中是考查程序设计能力的一个重要方面。

4.2.6 参考答案

```
#include <fstream>
#include <iostream>
#include <string>
using namespace std;
int main(int argc, char * argv[])
{
    ifstream cin("aaa.txt");
    int n;
    string s;
    cin>>n;
    for(int i=0;i<n;i++)
    {
        cin>>s;
        cout<<"String #"<<i+1<<endl;
        for(int j=0;j<s.size();j++)
        {
            cout<<(s[j]!='Z'?char(s[j]+1):'A');
        }
        cout<<endl;
        cout<<endl;//要产生一个空行
    }
    return 0;
}
```

4.2.7 汉语翻译

1. 题目

IBM 减 一

你或许听说过 Arthur C. Clarke 写的《2001：一次漫长的太空冒险旅行》这本书，或者 Stanley Kubrick 演的同名电影。它讲的是一艘太空飞船被从地球送到土星。飞船被智能计算机 HAL 控制，在这个漫长的飞行中，全体船员都将进入睡眠状态，只有两个人是醒着的。但是在飞行过程中 HAL 的行为越来越奇怪，甚至准备杀死所有船员。我们不告诉你这个故事的结尾，好让你自己去读这本书。

这部电影发行后很受大家的欢迎，许多人都在讨论"HAL"这个名字的实际含义。有些人认为 HAL 可能是"Heuristic ALgorithm"的缩写。但比较流行的解释是：把 HAL 中的每个字母按字母表里的顺序往后挪一个位置，就是 IBM。

可能有很多这样的首字母缩略语。你要编写一个程序来找出它们。

2. 输入描述

输入的第一行是一个整数 n，这个整数表示以下字符串的个数。接下来就是 n 行，每行上面有一个字符串，字符串是由至多 50 个大写字母构成。

3. 输出描述

对于每个输入的字符串，先输出该字符串的序号，以输出样例中的形式输出。然后输出一个字符串，该字符串的每个字符正好是输入字符串对应字母在字母表中的后一个位置上的字母，"Z"的后一个字母是"A"。

每个测试案例后面打印一个空行。

4. 输入样例

```
2
HAL
SWERC
```

5. 输出样例

```
String #1
IBM

String #2
TXFSD
```

4.3 Binary Numbers

4.3.1 链接地址

http://www.realoj.com/ 网上第 87 题

4.3.2 时空限制

Time Limit: 1000 ms   Resident Memory Limit: 1024 KB   Output Limit: 1024 B

4.3.3 题目内容

Given a positive integer n, print out the positions of all 1's in its binary representation. The position of the least significant bit is 0.

Example

The positions of 1's in the binary representation of 13 are 0, 2, 3.

Task

Write a program which for each data set: reads a positive integer n, computes the positions of 1's in the binary representation of n, writes the result.

Input

The first line of the input contains exactly one positive integer d equal to the number of data sets, $1 \leqslant d \leqslant 10$. The data sets follow. Each data set consists of exactly one line containing exactly one integer n, $1 \leqslant n \leqslant 10^6$.

Output

The output should consists of exactly *d* lines, one line for each data set. Line *i*, $1 \leq i \leq d$, should contain increasing sequence of integers separated by single spaces—the positions of 1's in the binary representation of the *i*-th input number.

Sample Input

1
13

Sample Output

0 2 3

4.3.4 题目来源

Central Europe 2001, Practice

4.3.5 解题思路

本题主要考察十进制到二进制的转换。十进制正整数转换为二进制整数的方法是不断除2取余，而被除数则不断除2取整，直到被除数变为0。把各位余数按相反的顺序连接起来，正好是该正整数的二进制表示。

本题要求打印一个正整数的二进制中1的位置，所以，只要把每位二进制位存入向量中即可，最后，再在向量中进行处理，这样很省事。

本题格式上，要求两位输出之间用一个空格隔开，而每行的最后不能有空格。这点要特别注意。

4.3.6 参考答案

```
#include <fstream>
#include <iostream>
#include <vector>
using namespace std;
int main(int argc, char * argv[])
{
    ifstream cin("aaa.txt");
    vector<int>v;
    int n,a;
    cin>>n;
    for(int i=0;i<n;i++)
    {
        cin>>a;
        v.clear();
        for(int j=a;j;j=j/2){
            v.push_back(j%2?1:0);
        }
```

```
            int p=0;//第一次输出
            for(int k=0;k<v.size();k++)
            {
                if(v[k]==1)
                {
                    if(p==0)cout<<k;
                    else cout<<" "<<k;
                    p=1;
                }
            }
            cout<<endl;
    }
    return 0;
}
```

4.3.7 汉语翻译

1. 题目

二 进 制 数

给出一个正整数 n，打印出它的二进制中所有 1 的位置。二进制中最低位的位置是 0。

举例

正整数 13 的二进制（1101）中 1 的位置是 0，2，3。

任务

编写一个程序来处理每个数据：

先读入一个正整数 n，再计算这个正整数的二进制中 1 的位置，打印结果。

2. 输入描述

输入数据的第一行是一个正整数 d，表示这个数据集中正整数的个数，$1 \leqslant d \leqslant 10$。数据集列在 d 的下面。每个数据集由一个整数 n（$1 \leqslant n \leqslant 10^6$）组成，且放在一行上。

3. 输出描述

输出应当有 d 行，一行是一个数据集处理的结果。

第 i 行，$1 \leqslant i \leqslant d$，应当包含该数据的二进制的 1 的位置，位置是按递增顺序排列，中间用一个空格隔开。

4. 输入样例

```
1
13
```

5. 输出样例

```
0 2 3
```

4.4 Encoding

4.4.1 链接地址

http://www.realoj.com/网上第 88 题

4.4.2 时空限制

Time Limit: 1000 ms   Resident Memory Limit: 1024 KB   Output Limit: 1024 B

4.4.3 题目内容

Given a string containing only "A" – "Z", we could encode it using the following method:

(1) Each sub-string containing k same characters should be encoded to "kX" where "X" is the only character in this sub-string.

(2) If the length of the sub-string is 1, "1" should be ignored.

Input

The first line contains an integer N ($1 \leq N \leq 100$) which indicates the number of test cases. The next N lines contain N strings. Each string consists of only "A" – "Z" and the length is less than 100.

Output

For each test case, output the encoded string in a line.

Sample Input

2
ABC
ABBCCC

Sample Output

ABC
A2B3C

4.4.4 题目来源

Zhejiang Provincial Programming Contest 2005 (Author: ZHANG, Zheng)

4.4.5 解题思路

本题是作一种简单的编码，把一个字符串中连续重复的字母从左到右写成 kX 的形式，如果 k 是 1，那么，1 就要省略。

本题的考察重点是对连续重复字符的处理，这种处理是字符串处理中常考的技巧。

由于 ACM 的判题系统只会比对输出结果的形式是否与要求的一致，不会考虑数据类型，所以，不要先把编码全部完成后一次性输出，而是要一边计算，一边输出结果，这点很重要。

4.4.6 参考答案

```cpp
#include <fstream>
#include <iostream>
#include <string>
using namespace std;
int main(int argc, char * argv[])
{
    ifstream cin("aaa.txt");
    string s,t;
    int n;
    cin>>n;
    for(int i=0;i<n;i++)
    {
        cin>>s;
        int c=0;
        t=s[0];
        int temp=0;
        for(int j=0;j<s.size();j++)
        {
            if(s[j]==t[0])
            {
                temp++;
                //如果已是最后一个，直接输出
                if(j==s.size()-1)
                {
                    if(temp==1)cout<<t[0];
                    else cout<<temp<<t[0];
                }
            }
            else
            {
                if(temp==1)cout<<t[0];
                else cout<<temp<<t[0];
                t[0]=s[j];
                temp=1;
                //如果已是最后一个，直接输出
                if(j==s.size()-1)
                {
                    if(temp==1)cout<<t[0];
                    else cout<<temp<<t[0];
                }
            }
        }
        cout<<endl;
        s="";
```

```
    }
    return 0;
}
```

4.4.7 汉语翻译

1. 题目

编　　码

给定一个只包含"A"~"Z"的字符串，我们使用下面的方法给它编码：
（1）将子字符串中的 k 个相同字符写成"kX"，X 是子串中的字符。
（2）如果子串的长度是1，那么"1"要忽略。

2. 输入描述

第一行包含一个正整数 N（$1 \leq N \leq 100$），代表测试案例的个数。下面 N 行包含 N 个字符串。每个字符串仅包含"A"~"Z"，且字符串的长度小于 100。

3. 输出描述

对于每个测试案例，输出它的编码在单独一行上。

4. 输入描述

   2
   ABC
   ABBCCC

5. 输出描述

   ABC
   A2B3C

4.5 Look and Say

4.5.1 链接地址

http://www.realoj.com/网上第89题

4.5.2 时空限制

Time Limit: 1000 ms   Resident Memory Limit: 1024 KB   Output Limit: 1024 B

4.5.3 题目内容

The look and say sequence is defined as follows. Start with any string of digits as the first element in the sequence. Each subsequent element is defined from the previous one by "verbally" describing the previous element. For example, the string 122344111 can be described as "one 1, two 2's, one 3, two 4's, three 1's". Therefore, the element that comes after 122344111 in the sequence is 1122132431. Similarly, the string 101 comes after 1111111111. Notice that it is generally not possible to uniquely identify the previous element of a particular element. For example, a string of 112213243 1's also yields 1122132431 as the next element.

Input

The input consists of a number of cases. The first line gives the number of cases to follow. Each case consists of a line of up to 1000 digits.

Output

For each test case, print the string that follows the given string.

Sample Input

```
3
122344111
1111111111
12345
```

Sample Output

```
1122132431
101
1112131415
```

4.5.4 题目来源

The 2007 ACM Rocky Mountain Programming Contest

4.5.5 解题思路

本题是处理重复子串的问题。虽然输入的都是数字,但应当把它们当成字符串来处理。由于本题时限一秒,每个字符串的长度多达 1 000 位,所以,不好的算法容易超时。

scanf 和 printf 所用的时间大大少于 cin 和 cout 所消耗的时间。由于本题需要频繁输出,采用 cout 则会超过一秒;而使用 printf 则不会超过一秒。这点是 ACM 竞赛中节约时间的常识。一般地,由于 cin 和 cout 调试很方便,所以调试期间使用它们,但是提交判题系统后,如果发现超时,可尝试将 cin 和 cout 改为 scanf 和 printf,看看是不是由于输入输出过于频繁而导致的。

4.5.6 参考答案

```cpp
#include <fstream>
#include <iostream>
#include <string>
using namespace std;
int main(int argc, char * argv[])
{
    ifstream cin("aaa.txt");
    string s,t;
```

```
    int n;
    cin>>n;
    for(int i=0;i<n;i++)
    {
        cin>>s;
        int c=0;
        t=s[0];
        int temp=0;
        for(int j=0;j<s.size();j++)
        {
            if(s[j]==t[0])
            {
                temp++;
                //如果已是最后一个,直接输出
                if(j==s.size()-1)
                {
                    //cout<<temp<<t[0];
                    printf("%d%c",temp,t[0]);
                }
            }
            else
            {
                //cout<<temp<<t[0];
                printf("%d%c",temp,t[0]);
                t[0]=s[j];
                temp=1;
                //如果已是最后一个,直接输出
                if(j==s.size()-1)
                {
                    //cout<<temp<<t[0];
                    printf("%d%c",temp,t[0]);
                }
            }
        }
        cout<<endl;
    }
    return 0;
}
```

4.5.7 汉语翻译

1. 题目

看 和 说

看和说系列定义如下:以数字字符串作为这个系列中的第一个元素。每个随后的元素是对它前面一个数字的口头描述。比如,字符串122344111可以被描述成"1个1,2个2,1个3,2个4,3个1"。因此,122344111应该写成1122132431。类似地,字符串1111111111可以写成101。

2. 输入描述

输入中包含多个测试案例。第一行是测试案例的个数。每个测试案例多达 1000 位。

3. 输出描述

对于每个测试案例，打印按题意要求的字符串。

4. 输入样例

```
3
122344111
1111111111
12345
```

5. 输出样例

```
1122132431
101
1112131415
```

4.6　Abbreviation

4.6.1　链接地址

http://www.realoj.com/网上第 90 题

4.6.2　时空限制

Time Limit: 1000 ms　　Resident Memory Limit: 1024 KB　　Output Limit: 1024 B

4.6.3　题目内容

When a Little White meets another Little White:

Little White A: (Surprised) !

Little White B: ?

Little White A: You Little White know "SHDC"? So unbelievable!

Little White B: You are Little White! Little White is you! What is "SHDC" you are talking about?

Little White A: Wait... I mean "Super Hard-disc Drive Cooler".

Little White B: I mean "Spade Heart Diamond Club"... Duck talks with chicken -_-//

Little White A: Duck... chicken... faint!

—quote from qmd of Spade6 in CC98 forum.

Sometimes, we write the abbreviation of a name. For example IBM is the abbreviation for International Business Machines. A name usually consists of one or more words. A word begins with a capital letter ("A"～"Z") and followed by zero or more lower-case letters ("a"～"z"). The abbreviation for a name is the word that consists of all the first letters of the words.

Now, you are given two names and asked to decide whether their abbreviations are the same.

Input

Standard input will contain multiple test cases. The first line of the input is a single integer T which is the number of test cases. And it will be followed by T consecutive test cases.

There are four lines for each case. The first line contains an integer N ($1 \leq N \leq 5$), indicating the number of words in the first name. The second line shows the first name. The third line contains an integer M ($1 \leq M \leq 5$), indicating the number of words in the second name. The fourth line shows the second name. Each name consists of several words separated by space. Length for every word is less than 10. The first letter for each word is always capital and the rest ones are lower-case.

Output

Results should be directed to standard output. The output of each test case should be a single line. If two names' abbreviations are the same, output "SAME", otherwise output "DIFFERENT".

Sample Input

```
3
4
Super Harddisc Drive Cooler
4
Spade Heart Diamond Club
3
Shen Guang Hao
3
Shuai Ge Hao
3
Cai Piao Ge
4
C P C S
```

Sample Output

```
SAME
SAME
DIFFERENT
```

4.6.4 题目来源

Zhejiang University Local Contest 2008(Author: HANG, Hang)

4.6.5 解题思路

本题是比较两个缩写词是否相同，而缩写词又是从一个包含多个单词的名字中合成的。每次读入一个单词，然后取出它的第一个字母，连接在字符串上，就组成了一个缩写词。

本题的难点在单词的读取控制上，对于输入数据的控制，是 ACM 竞赛中考查的一个重要方面。

另外，大家可以试试，使用 printf 输出比使用 cout 输出快很多。

4.6.6 参考答案

```
#include <fstream>
#include <iostream>
#include <string>
using namespace std;
int main(int argc, char * argv[])
{
    ifstream cin("aaa.txt");
    string s,ssa,ssb;
    int t,n,m;
    cin>>t;
    for(int i=0;i<t;i++)
    {
        cin>>n;
        for(int j=0;j<n;j++)
        {
            cin>>s;
            ssa=ssa + s[0];
        }
        cin>>m;
        for(int k=0;k<m;k++)
        {
            cin>>s;
            ssb=ssb + s[0];
        }
        if(ssa.compare(ssb)==0)//相等返回0,大于返回1,小于返回-1
            //cout<<"SAME"<<endl;
            printf("SAME\n");
        else
            //cout<<"DIFFERENT"<<endl;
            printf("DIFFERENT\n");
        ssa="";
        ssb="";
    }
    return 0;
}
```

4.6.7 汉语翻译

1. 题目

缩　　写

当 Little White A 遇到 Little White B：

Little White A：（吃惊）！

Little White B：？

Little White A：你知道"SHDC"？难以置信！

Little White B：你是 Little White！Little White 是你！你指的"SHDC"是什么？

Little White A：等等，我是指"Super Hard-disc Drive Cooler"。

Little White B：我以为是"Spade Heart Diamond Club"。真是鸭同鸡讲话-_-//

Little White A：鸭，鸡，晕！

——摘自 qmd of Spade6，CC98 论坛。

我们经常写缩写。如 IBM 是 International Business Machines 的缩写。一个名字通常包含多个单词。一个单词以一个大写字母打头（"A"～"Z"），后面不跟或跟多个小写字母（"a"～"z"）。缩写是由每个单词的首字母组成。

现在，给出两个名字，要求你说出这两个名字的缩写是否相同。

2. 输入描述

标准输入将包含多个测试案例。输入的第一行是一个整数 T，代表测试案例的个数。然后，是 T 组测试案例。

一组测试案例有四行。

第一行是一个整数 N（$1 \leq N \leq 5$），表示第一个名字中单词的个数。

第二行显示了第一个名字。

第三行是一个整数 M（$1 \leq M \leq 5$），表示第二个名字中单词的个数。

第四行显示了第二个名字。

每个名字由多个单词组成，单词间用一个空格分开。每个单词的长度少于 10 个。每个单词的第一个字母是大写字母，其余字母是小写字母。

3. 输出描述

结果应为标准输出。每个测试案例输出一行。如果两个名字的缩写相同，则输出"SAME"，否则，输出"DIFFERENT"。

4. 输入样例

```
3
4
Super Harddisc Drive Cooler
4
Spade Heart Diamond Club
3
Shen Guang Hao
3
Shuai Ge Hao
3
Cai Piao Ge
4
C P C S
```

5. 输出样例

```
SAME
SAME
DIFFERENT
```

4.7　The Seven Percent Solution

4.7.1　链接地址

http://www.realoj.com/网上第 91 题

4.7.2　时空限制

Time Limit: 1000 ms　Resident Memory Limit: 1024 KB　Output Limit: 1024 B

4.7.3　题目内容

Uniform Resource Identifiers (or URIs) are strings like *http://icpc.baylor.edu/icpc/*, *mailto: foo@bar.org*, *ftp://127.0.0.1/pub/linux*, or even just *readme.txt* that are used to identify a resource, usually on the Internet or a local computer.

Certain characters are reserved within URIs, and if a reserved character is part of an identifier then it must be *percent-encoded* by replacing it with a percent sign followed by two hexadecimal digits representing the ASCII code of the character. A table of seven reserved characters and their encodings is shown below. Your job is to write a program that can percent-encode a string of characters.

Character	Encoding
" " (space)	%20
"!" (exclamation point)	%21
"$" (dollar sign)	%24
"%" (percent sign)	%25
"(" (left parenthesis)	%28
")" (right parenthesis)	%29
"*" (asterisk)	%2a

Input

The input consists of one or more strings, each 1～79 characters long and on a line by itself, followed by a line containing only "#" that signals the end of the input. The character "#" is used only as an end-of-input marker and will not appear anywhere else in the input. A string may contain spaces, but not at the beginning or end of the string, and there will never be two or more consecutive spaces.

Output

For each input string, replace every occurrence of a reserved character in the table above by its percent-encoding, exactly as shown, and output the resulting string on a line by itself. Note that the percent-encoding for an asterisk is %2a (with a lowercase "a") rather than %2A (with an uppercase "A").

Sample Input

```
Happy Joy Joy!
http://icpc.baylor.edu/icpc/plain_vanilla
(**)
?
the 7% solution
#
```

Sample Output

```
Happy%20Joy%20Joy%21
http://icpc.baylor.edu/icpc/plain_vanilla
%28%2a%2a%29
?
the%207%25%20solution
```

4.7.4 题目来源

The 2007 ACM Mid-Central USA Programming Contest

4.7.5 解题思路

本题要求将一个字符串中的 7 个保留字符替换成 7 种百分比,由于一个字符串在一行上,一行中也存在若干个空格,所以,需要采用 cin.getline()方法一次读入一行,再去处理。

输出的时候特别要注意,由于%在 C 语言中是保留字符,使用 C 语言的 printf 函数输出,应当输出 "%%" 这个转义符。当然,如果采用 C++语言的 cout 输出,就不必考虑这个问题。这是两种输出的一个不同之处。

4.7.6 参考答案

```
#include <fstream>
#include <iostream>
#include <string>
using namespace std;
int main(int argc, char * argv[])
{
    ifstream cin("aaa.txt");
    string s;

    char ss[80];
    while(cin.getline(ss,80))
```

```
    {
        if(ss[0]=='#')break;
        s=ss;
        for(int i=0;i<s.size();i++)
        {
            //if(s[i]==' ')cout<<"%20";
            if(s[i]==' ')printf("%%20");
            //else if(s[i]=='!')cout<<"%21";
            else if(s[i]=='!')printf("%%21");
            //else if(s[i]=='$')cout<<"%24";
            else if(s[i]=='$')printf("%%24");
            //else if(s[i]=='%')cout<<"%25";
            else if(s[i]=='%')printf("%%25");
            //else if(s[i]=='(')cout<<"%28";
            else if(s[i]=='(')printf("%%28");
            //else if(s[i]==')')cout<<"%29";
            else if(s[i]==')')printf("%%29");
            //else if(s[i]=='*')cout<<"%2a";
            else if(s[i]=='*')printf("%%2a");
            //else cout<<s[i];
            else printf("%c",s[i]);
        }
        cout<<endl;
    }
    return 0;
}
```

4.7.7 汉语翻译

1. 题目

七个百分比编码

统一资源标识（或称 URI）是这样一些字符串"http://icpc.baylor.edu/icpc/"，"mailto:foo@bar.org"，"ftp://127.0.0.1/pub/linux"，甚至是只是"readme.txt"，它们通常是用来标识互联网上或本地计算机上的一个资源。

在 URI 中有一些保留字符，如果在 URI 中包含了这些保留字符，那么，这些保留字符就要被一个百分符号"%"加两位十进制数来代替。7 种百分比编码列在下表中。你的任务就是编制一个程序将保留字符用它们来替代。

字　符	编　码
" "（空格）	%20
"!"（感叹号）	%21
"$"（美元符）	%24
"%"（百分符）	%25
"("（左括号）	%28
")"（右括号）	%29
"*"（星号）	%2a

2. 输入描述

输入由一个或多个字符串组成，每个字符串由 1～79 个字符构成且在一行上，以"#"号结束。全部输入数据中，只在最后包含一个"#"号，不会出现在其他地方。一个字符串可能包含多个空格，但空格不出现在字符串的开头和结尾位置，也不会连续出现两个或两个以上的空格。

3. 输出描述

对于每个测试案例，使用表中的 7 种百分比将保留字符替代，每个测试案例输出一行。注意，星号的百分比替代符是%2a（这里是小写字母 a），而不是%2A（这里是大写字母 A）。

4. 输入样例

```
Happy Joy Joy!
http://icpc.baylor.edu/icpc/
plain_vanilla
(**)
?
the 7% solution
#
```

5. 输出样例

```
Happy%20Joy%20Joy%21
http://icpc.baylor.edu/icpc/
plain_vanilla
%28%2a%2a%29
?
the%207%25%20solution
```

4.8 Digital Roots

4.8.1 链接地址

http://www.realoj.com/网上第 92 题

4.8.2 时空限制

Time Limit: 1000 ms　　Resident Memory Limit: 1024 KB　　Output Limit: 1024 B

4.8.3 题目内容

Background

The digital root of a positive integer is found by summing the digits of the integer. If the resulting value is a single digit then that digit is the digital root. If the resulting value contains two or more digits, those digits are summed and the process is repeated. This is continued as long as necessary to obtain a single digit.

For example, consider the positive integer 24. Adding the 2 and the 4 yields a value of 6. Since 6 is a single digit, 6 is the digital root of 24. Now consider the positive integer 39. Adding the 3 and the 9 yields 12. Since 12 is not a single digit, the process must be repeated. Adding the 1 and the 2 yeilds 3, a single digit and also the digital root of 39.

Input

The input file will contain a list of positive integers, one per line. The end of the input will be indicated by an integer value of zero.

Output

For each integer in the input, output its digital root on a separate line of the output.

Example

Input

24
39
0

Output

6
3

4.8.4 题目来源

Greater New York 2000

4.8.5 解题思路

本题算法简单，但问题的关键在于，要把一个整数的各位分离出来，但如果采用取余的方法将各位分离出来，如果数据量太大，肯定会超时的。所以，这里把整数当成字符串读入，然后读取每个字符，那么就相当于将各位分离了，这种办法是比较省时的。

另外，对于 32 位无符号正整数，最大数为 $2^{32}-1=4\,294\,967\,296-1$（减去 1 是指减去 0 这种情况），所以，即便读入的整数是最大的数，即 9 999 999 999，第一次求所有位的和也是 90，因此，减少第一轮求和的计算量是问题的关键。

4.8.6 参考答案

```
#include <fstream>
#include <iostream>
#include <string>
using namespace std;
int main(int argc, char * argv[])
{
    ifstream cin("aaa.txt");
    string s;
    int sum;
```

```
        while(cin>>s)
        {
            if(s=="0")break;
            sum=0;
            //减少第一轮求和的计算是问题的关键
            for(int i=0;i<s.size();i++)
            {
                if(s[i]=='1')sum+=1;
                else if(s[i]=='2')sum+=2;
                else if(s[i]=='3')sum+=3;
                else if(s[i]=='4')sum+=4;
                else if(s[i]=='5')sum+=5;
                else if(s[i]=='6')sum+=6;
                else if(s[i]=='7')sum+=7;
                else if(s[i]=='8')sum+=8;
                else if(s[i]=='9')sum+=9;
            }
            //第二轮到第n轮求和
            while(1==1)
            {
                if(sum<10)
                {
                    //cout<<sum<<endl;
                    printf("%d\n",sum);
                    break;
                }
                else
                    sum=sum/10+sum%10;
            }
        }
        return 0;
    }
```

4.8.7 汉语翻译

1. 题目

数　字　根

背景

　　一个正整数的数字根是通过计算该整数的各位的和产生的。如果一个整数的个位和是一位整数，那么这个数字就是该整数的数字根。如果该整数的各位和是两位或多位整数，那么，就需要重复计算各位的和，直到获得一位整数。

　　例如，考虑正整数 24。把 2 与 4 相加得到 6。由于 6 是一个一位整数，所以，6 就是 24 的数字根。现在再来考虑正整数 39。3 与 9 相加等于 12。因为 12 不是一位整数，因而，需要重复处理。再把 1 加 2 得到 3，现在 3 已是一个一位整数了，那么 3 就是 39 的数字根。

2. 输入描述

输入文件包含一列正整数，一行一个。用 0 表示输入的结束。

3. 输出描述

对于输入的每个整数，在一行上单独输出它的数字根。

4. 举例

输入

24
39
0

输出

6
3

4.9 Box of Bricks

4.9.1 链接地址

http://www.realoj.com/网上第 93 题

4.9.2 时空限制

Time Limit: 1000 ms   Resident Memory Limit: 1024 KB   Output Limit: 1024 B

4.9.3 题目内容

Little Bob likes playing with his box of bricks. He puts the bricks one upon another and builds stacks of different height. "Look, I've built a wall!" he tells his older sister Alice. "Nah, you should make all stacks the same height. Then you would have a real wall." she retorts. After a little consideration, Bob sees that she is right. So he sets out to rearrange the bricks, one by one, such that all stacks are the same height afterwards. But since Bob is lazy he wants to do this with the minimum number of bricks moved. Can you help?

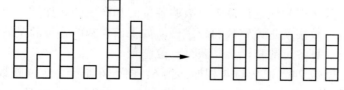

Input

The input consists of several data sets. Each set begins with a line containing the number n of stacks Bob has built. The next line contains n numbers, the heights h_i of the n stacks. You may assume $1 \leq n \leq 50$ and $1 \leq h_i \leq 100$. The total number of bricks will be divisible by the number of stacks. Thus, it is always possible to rearrange the bricks such that all stacks have the same height.

The input is terminated by a set starting with $n = 0$. This set should not be processed.

Output

For each set, first print the number of the set, as shown in the sample output. Then print the line "The minimum number of moves is k.", where k is the minimum number of bricks that have to be moved in order to make all the stacks the same height. Output a blank line after each set.

Sample Input

```
6
5 2 4 1 7 5
0
```

Sample Output

```
Set #1
The minimum number of moves is 5.
```

4.9.4　题目来源

Southwestern Europe 1997

4.9.5　解题思路

本题很有趣，把所有的砖盒移到同一高度，当然，就形成了一块四方形的砖墙了。每个砖盒的高度不同，需要计算出移动到同一高度所需的最小次数。

其实，先通过砖块的总数量除以砖墙的数量就得出了砖墙的最后高度，那么，用这个高度减去所有小于它的砖盒的高度，再把所有相减的结果加起来，就正好是需要移动的次数。

输出时，要注意输出格式应与题目要求的一致，否则，会出现输出格式错误。

4.9.6　参考答案

```cpp
#include <fstream>
#include <iostream>
#include <vector>
using namespace std;
int main(int argc, char * argv[])
{
    ifstream cin("aaa.txt");
    vector<int>v;
    int n,b,sum,avg,e,c=0;
    while(cin>>n)
    {
        if(n==0)break;
```

```
        c++;
        sum=0;
        e=0;
        v.clear();//清空向量
        for(int i=0;i<n;i++)
        {
            cin>>b;
            v.push_back(b);
            sum=sum+b;
        }
        avg=sum/n;
        for(int j=0;j<v.size();j++)
        {
            if(v[j]>avg)e=e+(v[j]-avg);
        }
        cout<<"Set #"<<c<<endl;
        cout<<"The minimum number of moves is "<<e<<"."<<endl;
        cout<<endl;//产生一空行
    }
    return 0;
}
```

4.9.7 汉语翻译

1. 题目

砖　盒

小 Bob 喜欢玩砖盒。他把砖块一个叠一个地堆成不同高度的盒子。"看，我堆了一堵墙！"他告诉他的大姐姐 Alice。"嗯，你该把所有的砖盒建得一样高。这样，你就建成了一堵真正的墙。"她回答。思索了一会，小 Bob 认为她是对的。然后，他就开始重新摆放砖盒，一块接一块，最后，所有的砖盒都一样高了。但 Bob 很懒，他想只移动最少数目的砖块来使所有的砖盒一样高。你能帮助他吗？

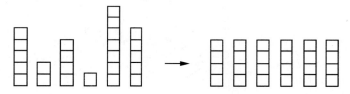

2. 输入描述

输入包含多个数据集合。每个数据集合的第一行是一个整数 n，表示该 Bob 建的砖盒个数。第二行则包含 n 个整数，每个整数代表每个砖盒的高度 h_i，表示该砖盒中有 h_i 块砖。假定 $1 \leqslant n \leqslant 50$，$1 \leqslant h_i \leqslant 100$。

砖块的总数目能被砖盒的数目整除。因此，一定能把各个砖盒移成相同的高度。

输入以 0 结束。不要处理 0。

3. 输出描述

对于每个数据集，先打印出该数据集的序号，请参考输出样例。然后打印这样一行"The minimum number of moves is k."。这里，k指把这些砖盒移到同一高度所需的最小次数。在每个数据集后输出一个空行。

4. 输入样例

```
6
5 2 4 1 7 5
0
```

5. 输出样例

```
Set #1
The minimum number of moves is 5.
```

4.10 Geometry Made Simple

4.10.1 链接地址

http://www.realoj.com/网上第94题

4.10.2 时空限制

Time Limit: 1000 ms　　Resident Memory Limit: 1024 KB　　Output Limit: 1024 B

4.10.3 题目内容

Mathematics can be so easy when you have a computer. Consider the following example. You probably know that in a right-angled triangle, the length of the three sides a, b, c (where c is the longest side, called the hypotenuse) satisfy the relation $a*a+b*b=c*c$. This is called Pythagora's Law.

Here we consider the problem of computing the length of the third side, if two are given.

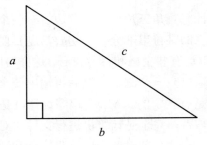

Input

The input contains the descriptions of several triangles. Each description consists of a line containing three integers a, b and c, giving the lengths of the respective sides of a right-angled triangle. Exactly one of the three numbers is equal to −1 (the "unknown" side), the others are positive (the "given" sides).

A description having $a=b=c=0$ terminates the input.

Output

For each triangle description in the input, first output the number of the triangle, as shown in the sample output. Then print "Impossible." If there is no right-angled triangle, that has the "given" side lengths. Otherwise output the length of the "unknown" side in the format "$s = l$", where s is the name of the unknown side (a, b or c), and l is its length. l must be printed exact to three digits to the right of the decimal point.

Print a blank line after each test case.

Sample Input

```
3 4 -1
-1 2 7
5 -1 3
0 0 0
```

Sample Output

```
Triangle #1
c = 5.000

Triangle #2
a = 6.708

Triangle #3
Impossible.
```

4.10.4 题目来源

Southwestern Europe 1997, Practice

4.10.5 解题思路

本题是使用勾股定理来计算直角三角形的一条边，有两个地方需要注意。

一个是输出格式，要输出的字符串部分，一定要与题目要求的一模一样，一个好办法是直接从网页中复制，因为里边有些空格和标点符号，这样可以保证字符串是一模一样的。另外一个格式是，本题要求输出完一个测试案例后再输出一个空行，而不是输出数据之间输出一个空行，这两种格式是有区别的。ACM 程序设计就是这样，追求精确，追求一致性。因为，输出结果的正确与否，完全是计算机进行文件比对的，所以，输出的结果一定要与答案文件里的样子一模一样才会判对，所以，我一直强调输出格式的重要性，再好的算法，如果输出格式不对，系统都会判错。

另一点就是要求输出小数点后三位有效数字，如果采用 cout 输出的话，先用

"cout.precision(3);"语句来指定输出精度是 3 位，再在输出前使用"cout<<fixed"来指定采用定点输出，也就是说，3 位精度是小数点后的三位。

4.10.6 参考答案

```cpp
#include <fstream>
#include <iostream>
#include <cmath>
using namespace std;
int main(int argc, char * argv[])
{
    ifstream cin("aaa.txt");
    int a,b,c,n;
    n=0;
    cout.precision(3);//有效数字三位
    while(cin>>a>>b>>c)
    {
        if(a==0&&b==0&&c==0)break;
        n++;
        if(a==-1)
        {
            if(c*c-b*b<=0)
            {
                cout<<"Triangle #"<<n<<endl;
                cout<<"Impossible."<<endl;
                cout<<endl;//产生空行
            }
            else
            {
                cout<<"Triangle #"<<n<<endl;
                //定点输出
                cout<<"a = "<<fixed<<pow(c*c-b*b,.5)<<endl;
                cout<<endl;//产生空行
            }
        }
        else if(b==-1)
        {
            if(c*c-a*a<=0)
            {
                cout<<"Triangle #"<<n<<endl;
                cout<<"Impossible."<<endl;
                cout<<endl;//产生空行
            }
            else
            {
                cout<<"Triangle #"<<n<<endl;
                //定点输出
                cout<<"b = "<<fixed<<pow(c*c-a*a,.5)<<endl;
```

```
                    cout<<endl;//产生空行
                }
            }
            else if(c==-1)
            {
                cout<<"Triangle #"<<n<<endl;
                //定点输出
                cout<<"c = "<<fixed<<pow(a*a+b*b,.5)<<endl;
                cout<<endl;//产生空行
            }
        }
        return 0;
    }
```

4.10.7 汉语翻译

1. 题目

简化几何计算

有了计算机后，数学计算变得如此简单。考虑下面这个例子。你可能清楚直角三角形中，三条边 a、b 和 c 的长度（这里 c 是最长的一条，叫做斜边）具有这个关系 $a*a + b*b = c*c$。这个公式称为勾股定理。

现在给出直角三角形的两条边，要求计算出第三条边。

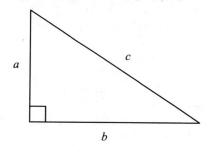

2. 输入描述

输入包含多个三角形的描述。每个描述在一行上，包括三个整数 a、b 和 c，表示三角形的三条边长。三个数中有一个是-1（未知边），其余两个是正数（已知的边）。
$a=b=c=0$ 表示输入的结束。

3. 输出描述

对于输入数据中的每组三角形描述，先输出三角形的序号，请参考输出样例。如果不存在这样的直角三角形，再输出"Impossible."。否则，直接输出未知边的长度，形式是"s = l"，这里 s 是未知边的名称（a、b 或 c），l 是指它的长度。l 必须精确到小数点后 3 位数字。

每一组测试案例后输出一个空行。

4. 输入样例

```
3 4 -1
-1 2 7
5 -1 3
0 0 0
```

5. 输出样例

```
Triangle #1
c = 5.000

Triangle #2
a = 6.708

Triangle #3
Impossible.
```

4.11 Reverse Text

4.11.1 链接地址

http://www.realoj.com/网上第 95 题

4.11.2 时空限制

Time Limit: 1000 ms   Resident Memory Limit: 1024 KB   Output Limit: 1024 B

4.11.3 题目内容

In most languages, text is written from left to right. However, there are other languages where text is read and written from right to left. As a first step towards a program that automatically translates from a left-to-right language into a right-to-left language and back, you are to write a program that changes the direction of a given text.

Input Specification

The input contains several test cases. The first line contains an integer specifying the number of test cases. Each test case consists of a single line of text which contains at most 70 characters. However, the newline character at the end of each line is not considered to be part of the line.

Output Specification

For each test case, print a line containing the characters of the input line in reverse order.

Sample Input

```
3
Frankly, I don't think we'll make much
```

```
money out of this scheme.
madam I'm adam
```

Sample Output

```
hcum ekam ll'ew kniht t'nod I ,ylknarF
.emehcs siht fo tuo yenom
mada m'I madam
```

4.11.4 题目来源

Southwestern Europe 1996, Practice

4.11.5 解题思路

本题要求将一行文本从左到右完全翻转，即镜像。采用 reverse 算法，即可一次性将一个字符串 string 从头到尾翻转，这就是泛型编程的强大功能。在比赛中，如果采用 C 语言写个循环来处理这样的事情，那需要比较多的时间去编程，还可能出错，显然，没有任何竞争力。

本题采用 cin.getline() 进行行输入，由于数据比较规范，所以，第一行的整数，可以忽略掉。程序只要一行一行往下读就行了，直到读不出数据就终止。这点，给我们一些启示，输入数据中的所有数据并不是都对我们有用，有些数据，完全可以忽略掉，这就是解题上的灵活性。

4.11.6 参考答案

```
#include <fstream>
#include <iostream>
#include <algorithm>
#include <string>
using namespace std;
int main(int argc, char * argv[])
{
    ifstream cin("aaa.txt");
    string s;
    char ss[80];
    int n=0;
    while(cin.getline(ss,80))
    {
        if(n==0)n=1;//跳过第一行输入
        else
        {
            s=ss;
            reverse(s.begin(),s.end());
            cout<<s<<endl;
        }
    }
}
```

```
        return 0;
    }
```

4.11.7 汉语翻译

1. 题目

反 转 文 本

在多数语言中，文本都是从左写到右。然而，有些语言中，文本是从右读写到左。现在，你将来编写一个程序，把一段从左到右书写的文本自动转换为从右写到左的文本。

2. 输入详细情况

输入包含多个测试案例。第一行是一个整数，代表测试案例的数目。每个测试案例由一行组成，至多 70 个字符。但是，每行末尾的换行符不作为该测试案例的字符。

3. 输出详细情况

对于每个测试案例，在一行上打印出输入行的反向文本。

4. 输入样例

```
3
Frankly, I don't think we'll make much
money out of this scheme.
madam I'm adam
```

5. 输出样例

```
hcum ekam ll'ew kniht t'nod I ,ylknarF
.emehcs siht fo tuo yenom
mada m'I madam
```

4.12 Word Reversal

4.12.1 链接地址

http://www.realoj.com/网上第 96 题

4.12.2 时空限制

Time Limit: 1000 ms   Resident Memory Limit: 1024 KB   Output Limit: 1024 B

4.12.3 题目内容

For each list of words, output a line with each word reversed without changing the order of the words.

This problem contains multiple test cases!

The first line of a multiple input is an integer N, then a blank line followed by N input blocks. Each input block is in the format indicated in the problem description. There is a blank line between input blocks.

The output format consists of N output blocks. There is a blank line between output blocks.

Input

You will be given a number of test cases. The first line contains a positive integer indicating the number of cases to follow. Each case is given on a line containing a list of words separated by one space, and each word contains only uppercase and lowercase letters.

Output

For each test case, print the output on one line.

Sample Input

```
1

3
I am happy today
To be or not to be
I want to win the practice contest
```

Sample Output

```
I ma yppah yadot
oT eb ro ton ot eb
I tnaw ot niw eht ecitcarp tsetnoc
```

4.12.4 题目来源

East Central North America 1999, Practice

4.12.5 解题思路

本题是将一行中的单词一个一个反转过来，但单词的位置不变。看起来比较简单，但由于存在多个数据块（block），所以，程序控制上比较难。

本题最容易出现呈现错误（Present Error），即答案是对的，而格式不对。一定要注意，输出的每个 block 之间有一空行，最后一个 block 后面没有空行。

4.12.6 参考答案

```cpp
#include <fstream>
#include <iostream>
#include <string>
using namespace std;
int main(int argc, char * argv[])
{
    ifstream cin("aaa.txt");
```

```
    string s,ss;

    char ch[81];
    cin.getline(ch,81);//不处理块数量
    cin.getline(ch,81);//不处理第一块的行数量
    int f=1;//标记是否是第一行
    while(cin.getline(ch,81))
    {
        s=ch;
        //找到返回从 0 开始的位置，否则返回 4294967295
        if(s.find(' ')==4294967295 && f!=1)//是行数量又不是第一块
        {
            cin.getline(ch,81);//不处理该块的行数量
            cout<<endl;
        }
        else f=0;
        //行中有空格存在，说明是语句行
        if(s.find(' ')!=4294967295)
        {
            for(unsigned int i=0;i<s.size();i++)
            {
                if(s[i]!=' ')
                    ss=s[i]+ss;//放到首位
                else
                {
                    cout<<ss<<" ";
                    ss="";
                }
            }
            cout<<ss<<endl;
            ss="";
        }
    }
    return 0;
}
```

4.12.7 汉语翻译

1. 题目

反 转 单 词

对于一串单词，直接把它们输出在一行上，要把每个单词反转，但每个单词的位置不要改变。

本程序包含多个测试案例！

输入数据的第一行是一个整数 N，然后是一空行，后面跟着 N 个数据块。每个数据块的格式在程序描述中说明了。数据块中有一空行。

输出格式由 N 个输出块组成。每个输出块间有一空行。

2. 输入描述

现在给你多组测试案例。第一行是一个正整数，表示接下来的测试案例的个数。每个测试案例是一行，包含一串单词，中间用一个空格隔开，每个单词仅包含大写和小写字母。

3. 输出描述

每个测试案例都打印在一行上。

4. 输入样例

```
1

3
I am happy today
To be or not to be
I want to win the practice contest
```

5. 输出样例

```
I ma yppah yadot
oT eb ro ton ot eb
I tnaw ot niw eht ecitcarp tsetnoc
```

4.13 A Simple Question of Chemistry

4.13.1 链接地址

http://www.realoj.com/网上第 97 题

4.13.2 时空限制

Time Limit: 1000 ms   Resident Memory Limit: 1024 KB   Output Limit: 1024 B

4.13.3 题目内容

Your chemistry lab instructor is a very enthusiastic graduate student who clearly has forgotten what their undergraduate Chemistry 101 lab experience was like. Your instructor has come up with the brilliant idea that you will monitor the temperature of your mixture every minute for the entire lab. You will then plot the rate of change for the entire duration of the lab.

Being a promising computer scientist, you know you can automate part of this procedure, so you are writing a program you can run on your laptop during chemistry labs. (Laptops are only occasionally dissolved by the chemicals used in such labs) You will write a program that will let you enter in each temperature as you observe it. The program will then calculate the difference between this temperature and the previous one, and print out the difference. Then you can feed this input into a simple graphing program and finish your plot before you leave the chemistry lab.

Input

The input is a series of temperatures, one per line, ranging from -10 to 200. The temperatures may be specified up to two decimal places. After the final observation, the number 999 will indicate the end of the input data stream. All data sets will have at least two temperature observations.

Output

Your program should output a series of differences between each temperature and the previous temperature. There is one fewer difference observed than the number of temperature observations (output nothing for the first temperature). Differences are always output to two decimal points, with no leading zeroes (except for the ones place for a number less than 1, such as 0.01) or spaces.

After the final output, print a line with "End of Output".

Sample Input

```
10.0
12.05
30.25
20
999
```

Sample Output

```
2.05
18.20
-10.25
End of Output
```

4.13.4 题目来源

Mid-Atlantic USA 2003

4.13.5 解题思路

本题很简单，只要将后面一个数字减前面一个数字即可，主要考查对小数点的处理。

本题要求频繁输出，故输出上用的时间会多些。采用 printf 输出比采用 cout 输出来得快。

采用 printf 控制小数点位数形式为：`printf("%*.*f",a);`

采用 cout 输出小数点后两位形式为：

```
cout.precision(2);   //设置精度为 2 位
cout<<fixed<<b-a;    //fixed 表示定点输出
```

4.13.6 参考答案

```
#include <fstream>
#include <iostream>
```

```cpp
using namespace std;
int main(int argc, char * argv[])
{
    ifstream cin("aaa.txt");
    double a,b;
    cin>>a;
    //cout.precision(2);
    while(cin>>b)
    {
        if(b==999)
        {
            //cout<<"End of Output"<<endl;
            printf("End of Output\n");
            break;
        }
        //cout<<fixed<<b-a<<endl;
        printf("%.2f\n",(b-a));
        a=b;
    }
    return 0;
}
```

4.13.7 汉语翻译

1. 题目

一个简单的化学问题

你化学实验室的助手是一个非常热心的研究生，他很显然忘记了他们大学时期 101 次在化学实验室的经历。他想出了一个好点子，你可以在整个实验期间每分钟观察一次混合物的温度，这样，你就可以画出整个实验期间温度的改变比率。

作为一个有前途的计算机科学家，你知道你能够将上述部分过程自动化，所以，你正编写一个运行在笔记本上的程序供化学实验期间使用。（在这样的化学实验室中，膝上电脑很少会发生故障）你编写的程序可以让你输入整个实验过程中观察到的温度。程序会自动计算出当前温度与前一个温度之间的差，然后，把这个差打印出来。最后，你可以将这些差填到一个简单的图表程序里，使得在你离开实验室前完成画图工作。

2. 输入描述

输入是一系列的温度，一行一个，范围在-10～200 之间。温度至多有两位小数。观察结束后，输入"999"表示整个数据输入的结束。所有数据集至少包含两个温度。

3. 输出描述

你的程序将输出一系列的差，差是由每个温度减去头一个温度而得来的。观察到的温度基本上与实际温度相差不大（第一个温度不要处理）。输出的数据要保留小数点后两位数字，一个数开头的 0 要去掉（除非是小于 1 的数，当然，整数部分是要有一个 0 的，比如 0.01），一个数开头也不要有空格。

全部输出完成后，打印一行"End of Output"。

4. 输入样例

```
10.0
12.05
30.25
20
999
```

5. 输出样例

```
2.05
18.20
-10.25
End of Output
```

4.14 Adding Reversed Numbers

4.14.1 链接地址

http://www.realoj.com/网上第 98 题

4.14.2 时空限制

Time Limit: 1000 ms　　Resident Memory Limit: 1024 KB　　Output Limit: 1024 B

4.14.3 题目内容

The Antique Comedians of Malidinesia prefer comedies to tragedies. Unfortunately, most of the ancient plays are tragedies. Therefore the dramatic advisor of ACM has decided to transfigure some tragedies into comedies. Obviously, this work is very hard because the basic sense of the play must be kept intact, although all the things change to their opposites. For example the numbers: if any number appears in the tragedy, it must be converted to its reversed form before being accepted into the comedy play.

Reversed number is a number written in arabic numerals but the order of digits is reversed. The first digit becomes last and vice versa. For example, if the main hero had 1245 strawberries in the tragedy, he has 5421 of them now. Note that all the leading zeros are omitted. That means if the number ends with a zero, the zero is lost by reversing (e.g. 1200 gives 21). Also note that the reversed number never has any trailing zeros.

ACM needs to calculate with reversed numbers. Your task is to add two reversed numbers and output their reversed sum. Of course, the result is not unique because any particular number is a reversed form of several numbers (e.g. 21 could be 12, 120 or 1200 before reversing). Thus we must assume that no zeros were lost by reversing (e.g. assume that the original number was 12).

Input

The input consists of *N* cases. The first line of the input contains only positive integer *N*. Then follow the cases. Each case consists of exactly one line with two positive integers separated by space. These are the reversed numbers you are to add.

Output

For each case, print exactly one line containing only one integer — the reversed sum of two reversed numbers. Omit any leading zeros in the output.

Sample Input

```
3
24 1
4358 754
305 794
```

Sample Output

```
34
1998
1
```

4.14.4 题目来源

Central Europe 1998

4.14.5 解题思路

本题要求先把两个数反转过来后，再相加。相加后，再把结果反转过来，如果结果中包含前导零，则要删除前导零。

这样计算可以避免数字的多次反转操作：把两个数高位对齐再相加，相加时，从最高位加到最低位，并向低位进位。相加完成后，把末尾跟着的零全部删除，最后，把该数字反转过来，就是题目要求的输出结果。

本题把数字当成字符串看待，这样，一位一位便于区分，而且在相加的时候，要设置一个进位标志。处理时，要特别仔细。本题的解法同样适用于两个超长整数的相加运算。

4.14.6 参考答案

```cpp
#include <fstream>
#include <iostream>
#include <vector>
#include <string>
using namespace std;
int main(int argc, char * argv[])
{
    ifstream cin("aaa.txt");
    string sa,sb,st;
    vector<int>v;
```

```
int a,b,sum;
int p=0;//进位标志
int u=0;//输出标志
cin>>sa;//不处理案例数量
while(cin>>sa>>sb)
{
    if(sa.size()<sb.size())
    {
        st=sa;
        sa=sb;
        sb=st;
    }
    for(int i=0;i<sa.size();i++)
    {
        if(sa[i]=='0')a=0;
        else if(sa[i]=='1')a=1;
        else if(sa[i]=='2')a=2;
        else if(sa[i]=='3')a=3;
        else if(sa[i]=='4')a=4;
        else if(sa[i]=='5')a=5;
        else if(sa[i]=='6')a=6;
        else if(sa[i]=='7')a=7;
        else if(sa[i]=='8')a=8;
        else if(sa[i]=='9')a=9;

        if(i>=sb.size())b=0;
        else if(sb[i]=='0')b=0;
        else if(sb[i]=='1')b=1;
        else if(sb[i]=='2')b=2;
        else if(sb[i]=='3')b=3;
        else if(sb[i]=='4')b=4;
        else if(sb[i]=='5')b=5;
        else if(sb[i]=='6')b=6;
        else if(sb[i]=='7')b=7;
        else if(sb[i]=='8')b=8;
        else if(sb[i]=='9')b=9;

        sum=a+b+p;//两位求和再加上进位p
        p=sum/10;//求本次的进位,即十位值
        v.push_back(sum%10);//个位
    }
    if(p==1)v.push_back(1);
    //去掉末尾的所有0
    while(1)
    {
        vector<int>::iterator it=v.end()-1;
        if(*it==0)v.erase(it);
        else break;
    }
    for(int j=0;j<v.size();j++)
```

```
            {
                if(u==0 && v[j]!=0)u=1;
                if(u==1)cout<<v[j];
            }
            cout<<endl;
            v.clear();
            u=0;
            p=0;
        }
        return 0;
    }
```

4.14.7 汉语翻译

1. 题目

反转的数字相加

古代喜剧演员 Malidinesia 爱演喜剧而不爱演悲剧。不巧的是，绝大多数古代戏剧都是悲剧。因此 ACM 戏剧顾问决定把一些悲剧改写成喜剧。显然，这将是一项艰难的工作。虽然所有内容都可以改成相对的内容，但戏剧的根本首先是要保持意义的完整性。以数字为例，悲剧中出现数字时，就必须把数字反转。

反转的数字是指把阿拉伯数字反过来。也就是说，数字的第一位变成最后一位，其他位亦如此。比如，在悲剧中主角有 1 245 颗草莓，现在，他就拥有 5 421 颗了。注意，所有的前导零都要被删除。也就是说，如果数字是以 0 结尾的，那么，反转后，0 就没有了（比如，1 200 反转后就是 21）。另外，要注意反转数字的尾部不可能还有 0。

ACM 需要计算反转的数字。你的任务是把两个反转的数字相加，并输出它们的和的反转数。当然，结果不是唯一的，因为任何一个特定的数字都是多个数字反转的形式（比如，21 可能是 12、120 或 1 200 反转的结果）。因此，我们必须假定在反转的时候不会发生丢失 0 的情况（例如，假定数字是 12）。

2. 输入描述

输入包含 N 个测试案例。输入数据的第一行是一个正整数 N。接下去是测试数据。每个测试案例由一行组成，包括两个正整数，中间用空格隔开。这两个数就是你要进行加法运算的反转数字。

3. 输出描述

对于每个测试案例，一行只打印一个整数——这个数就是两个反转数的和的反转数。要删除所有的前导零。

4. 输入样例

```
3
24 1
4358 754
305 794
```

5. 输出样例

```
34
1998
1
```

4.15 Image Transformation

4.15.1 链接地址

http://www.realoj.com/网上第 99 题

4.15.2 时空限制

Time Limit: 1000 ms   Resident Memory Limit: 1024 KB   Output Limit: 1024 B

4.15.3 题目内容

The image stored on a computer can be represented as a matrix of pixels. In the RGB (Red-Green-Blue) color system, a pixel can be described as a triplex integer numbers. That is, the color of a pixel is in the format "r g b" where r, g and b are integers ranging from 0 to 255 (inclusive) which represent the Red, Green and Blue level of that pixel.

Sometimes however, we may need a gray picture instead of a colorful one. One of the simplest way to transform a RGB picture into gray: for each pixel, we set the Red, Green and Blue level to a same value which is usually the average of the Red, Green and Blue level of that pixel (that is (r + g + b)/3, here we assume that the sum of r, g and b is always dividable by 3).

You decide to write a program to test the effectiveness of this method.

Input

The input contains multiple test cases!

Each test case begins with two integer numbers N and M ($1 \leq N, M \leq 100$) meaning the height and width of the picture, then three $N * M$ matrices follow; respectively represent the Red, Green and Blue level of each pixel.

A line with $N = 0$ and $M = 0$ signals the end of the input, which should not be proceed.

Output

For each test case, output "Case #:" first. "#" is the number of the case, which starts from 1. Then output a matrix of $N * M$ integers which describe the gray levels of the pixels in the resultant grayed picture. There should be N lines with M integers separated by a comma.

Sample Input

```
2 2
1 4
6 9
```

```
2 5
7 10
3 6
8 11
2 3
0 1 2
3 4 2
0 1 2
3 4 3
0 1 2
3 4 4
0 0
```

Sample Output

```
Case 1:
2,5
7,10
Case 2:
0,1,2
3,4,3
```

4.15.4 题目来源

Zhejiang Provincial Programming Contest 2007 (Author: ZHOU, Yuan)

4.15.5 解题思路

本题看似简单，实际上很不容易。简单是指计算方法十分简单，只需使用(r+g+b)/3即可得出灰度值；而难在理解题意上，输入数据中，先列出了图像中所有像素的红色值，再列出所有像素（$N*M$ 个）的绿色值，最后列出的是所有像素的蓝色值。很多同学都认为每三个数据分别是指一个像素的红、绿和蓝值，这样理解就错了。遇到这样的问题，大家可以事先从输入、输出样例中推断。

4.15.6 参考答案

```
#include <fstream>
#include <iostream>
#include <vector>
using namespace std;
int main(int argc, char * argv[])
{
    ifstream cin("aaa.txt");
    vector<int>r;
    vector<int>g;
    vector<int>b;
    int n,m;
    int rr,gg,bb;
```

```
    int w=0;//样例数量
    while(cin>>n>>m)
    {
        r.clear();
        g.clear();
        b.clear();
        w++;
        if(n==0 && m==0)break;
        for(int i=0;i<n*m;i++)
        {
            cin>>rr;
            r.push_back(rr);
        }
        for(int j=0;j<n*m;j++)
        {
            cin>>gg;
            g.push_back(gg);
        }
        for(int k=0;k<n*m;k++)
        {
            cin>>bb;
            b.push_back(bb);
        }
        cout<<"Case "<<w<<":"<<endl;
        for(int p=0;p<n*m;p++)
        {
            cout<<(r[p]+g[p]+b[p])/3;
            if((p+1)%m==0)cout<<endl;
            else cout<<",";
        }
    }
    return 0;
}
```

4.15.7 汉语翻译

1. 题目

图 像 转 换

图像是以像素矩阵的形式存储在计算机里。在红绿蓝三色系统中,一个像素采用三个整数来表示。也就是说,一个像素的颜色以"r g b"的格式表示,这里,r,g 和 b 是 0~255 之间(包括 0 和 255)的整数,分别代表该像素红、绿和蓝的程度。

然而有时候,我们需要灰度图像而不是彩色图像。把 RGB 图像转换为灰度图像的一种最简便的方法是:把一个像素的红、绿和蓝的值都设置为一个相同的值(即(r+g+b)/3,这里假定(r+g+b)总能被 3 整除)。

你决定编写一个程序来测试这种方法的有效性。

2. 输入描述

输入包含多个测试案例！

每个测试案例以两个整数 N 和 M（$1 \leq N, M \leq 100$）打头，表示图像的高度和宽度，接下来，是三个 $N * M$ 矩阵，分别代表每个像素的红、绿和蓝的值。

在一行上，$N=0$ 和 $M=0$ 表示输入的结束，这行数据不需处理。

3. 输出描述

对于每个测试案例，先输出"Case #:"。"#"是测试案例的序号，从 1 开始计数。然后，输出一个 $N*M$ 的矩阵，它描述了最后的灰度图像每个像素的灰度值。应当有 N 行，每行上有 M 个整数，其间用逗号隔开。

4. 输入样例

```
2 2
1 4
6 9
2 5
7 10
3 6
8 11
2 3
0 1 2
3 4 2
0 1 2
3 4 3
0 1 2
3 4 4
0 0
```

5. 输出样例

```
Case 1:
2,5
7,10
Case 2:
0,1,2
3,4,3
```

4.16 Beautiful Meadow

4.16.1 链接地址

http://www.realoj.com/ 网上第 100 题

4.16.2 时空限制

Time Limit: 1000 ms   Resident Memory Limit: 1024 KB   Output Limit: 1024 B

4.16.3 题目内容

Tom has a meadow in his garden. He divides it into $N*M$ squares. Initially all the squares were covered with grass. He mowed down the grass on some of the squares and thinks the meadow is beautiful if and only if

(1) Not all squares are covered with grass.

(2) No two mowed squares are adjacent.

Two squares are adjacent if they share an edge. Here comes the problem: Is Tom's meadow beautiful now?

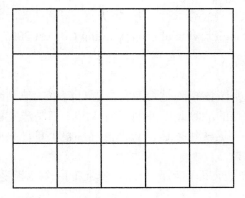

Tom's Meadow

Input

The input contains multiple test cases!

Each test case starts with a line containing two integers N, M ($1 \leqslant N$, $M \leqslant 10$) separated by a space. There follows the description of Tom's Meadow. There're N lines each consisting of M integers separated by a space. 0 (zero) means the corresponding position of the meadow is mowed and 1 (one) means the square is covered by grass.

A line with $N=0$ and $M=0$ signals the end of the input, which should not be processed.

Output

One line for each test case.

Output "Yes" (without quotations) if the meadow is beautiful, otherwise "No" (without quotations).

Sample Input

```
2 2
1 0
0 1
2 2
1 1
```

```
0 0
2 3
1 1 1
1 1 1
0 0
```

Sample Output

```
Yes
No
No
```

4.16.4 题目来源

Problem Source: Zhejiang Provincial Programming Contest 2007 (Author: CAO, Peng)

4.16.5 解题思路

本题针对一个矩阵，判断该矩阵是否漂亮，题目简单，但判断的逻辑需要好好理顺。下面两种情况是不漂亮的，除此以外的任何情况，都是漂亮的：

（1）全是 1，不漂亮。这就包括这样一种情况，整个草坪只有一个方块，如果方块的草没有被剪去，那就不漂亮。

（2）两个方块共一条边，即前后两个元素或上下两个元素是 00。

4.16.6 参考答案

```cpp
#include <fstream>
#include <iostream>
using namespace std;
int main(int argc, char * argv[])
{
    ifstream cin("aaa.txt");
    int p[10][10];
    int n,m;
    int i,j,k;
    int flag=1;//1 表示没锄过草
    while(cin>>n>>m)
    {
        if(n==0 && m==0)break;//读数据结束
        flag =1;//先标记没有锄过草
        //读入一个矩阵
        for(i=0;i<n;i++)
        {
            for(j=0;j<m;j++)
            {
                cin>>p[i][j];
                if(p[i][j]==0)flag=0;//修剪过
            }
```

```
        }
        //开始判断
        //第一种情况：没修剪过，即全是1，则不漂亮
        if(flag==1)
        {
            cout<<"No"<<endl;
            continue;
        }
        //第二种情况：剪的两块共一条边，即前后两个元素或上下两个元素是00，不漂亮
        //判断第0行
        for(k=1;k<m;k++)
        {
            if(p[0][k]==0 && p[0][k-1]==0)
            {
                cout<<"No"<<endl;
                goto RL;
            }
        }
        //判断第1~n-1行
        for(i=1;i<n;i++)
        {
            for(int j=0;j<m;j++)
            {
                //本行与上一行同一列元素都是0吗
                if(p[i][j]==0 && p[i-1][j]==0)
                {
                    cout<<"No"<<endl;
                    goto RL;//结束
                }
                //本行中当前元素和前一元素都是0吗
                if(j!=0)
                {
                    if(p[i][j]==0 && p[i][j-1]==0)
                    {
                        cout<<"No"<<endl;
                        goto RL;//结束
                    }
                }
            }
        }
        cout<<"Yes"<<endl;//漂亮
        continue;
RL:
        continue;
    }
    return 0;
}
```

4.16.7 汉语翻译

1. 题目

漂亮的草坪

Tom 的花园里有一块草坪。他把它分成 $N*M$ 的方块。一开始，所有的方块上都长着草。他剪去一些方块上的草，并认为仅符合这两个条件这块草坪才算漂亮：

（1）不是所有的方块上都长着草。
（2）两块剪去草的方块不能相连。

如果两块方块共一条边，那么就算是连着的。现在问题来了：Tom 的草坪漂亮吗？

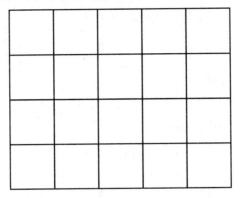

Tom 的草坪

2. 输入描述

输入包含多个测试案例！

每个测试案例以包含两个整数 N, M（$1 \leq N, M \leq 10$）开始，这两个数占一行，中间用空格隔开。接下来是 Tom 的草坪的描述。有 N 行，每行有 M 个整数，用空格隔开。0 表示该方块被剪去的草，1 表示该方块长着草。

一行上如果是 $N=0$ 且 $M=0$，表示输入的结束，这行不要处理。

3. 输出描述

一个测试案例输出一行。

如果这块草坪是漂亮的，输出"Yes"，否则输出"No"（不包括引号）。

4. 输入样例

```
2 2
1 0
0 1
2 2
1 1
0 0
```

```
2 3
1 1 1
1 1 1
0 0
```

5. 输出样例

```
Yes
No
No
```

4.17　DNA Sorting

4.17.1　链接地址

http://www.realoj.com/网上第 101 题

4.17.2　时空限制

Time Limit: 1000 ms　Resident Memory Limit: 1024 KB　Output Limit: 1024 B

4.17.3　题目内容

One measure of "unsortedness" in a sequence is the number of pairs of entries that are out of order with respect to each other. For instance, in the letter sequence "DAABEC", this measure is 5, since D is greater than four letters to its right and E is greater than one letter to its right. This measure is called the number of inversions in the sequence. The sequence "AACEDGG" has only one inversion (E and D)—it is nearly sorted—while the sequence "ZWQM" has 6 inversions (it is as unsorted as can be—exactly the reverse of sorted).

You are responsible for cataloguing a sequence of DNA strings (sequences containing only the four letters A, C, G, and T). However, you want to catalog them, not in alphabetical order, but rather in order of "sortedness", from "most sorted" to "least sorted". All the strings are of the same length.

This problem contains multiple test cases!

The first line of a multiple input is an integer N, then a blank line followed by N input blocks. Each input block is in the format indicated in the problem description. There is a blank line between input blocks.

The output format consists of N output blocks. There is a blank line between output blocks.

Input

The first line contains two integers: a positive integer n ($0 < n \leq 50$) giving the length of the strings; and a positive integer m ($1 < m \leq 100$) giving the number of strings. These are followed by m lines, each containing a string of length n.

Output

Output the list of input strings, arranged from "most sorted" to "least sorted". If two or more strings are equally sorted, list them in the same order they are in the input file.

Sample Input

```
1
10 6
AACATGAAGG
TTTTGGCCAA
TTTGGCCAAA
GATCAGATTT
CCCGGGGGGA
ATCGATGCAT
```

Sample Output

```
CCCGGGGGGA
AACATGAAGG
GATCAGATTT
ATCGATGCAT
TTTTGGCCAA
TTTGGCCAAA
```

4.17.4 题目来源

East Central North America 1998

4.17.5 解题思路

本题是字符串排序问题，但不是按字符串的 ASCII 大小排序，而是按每个字符串的倒位数量来排序。

所谓倒位数量，就是该字符串中，每个字符与它右边的字符相比，逆序的次数的总和。明确了这个计算方法，那么，程序是比较好实现的。

排序时，如果两行字符串移动的次数相同（即倒位总量相等），那么按原先固有的位置排序；输出时，每块之间需要产生一行空行。

4.17.6 参考答案

```cpp
#include <fstream>
#include <iostream>
#include <string>
#include <vector>
#include <algorithm>
using namespace std;
//自定义比较函数
bool comp(const string &s1,const string &s2)
```

```cpp
{
    int i,j,k,m,n;
    int c1=0,c2=0;
    //计算s1需要移动的次数c1
    for(i=0;i<s1.size();i++)
    {
        for(j=i+1;j<s1.size();j++)
        {
            if(s1[i]>s1[j])
                c1++;
        }
    }
    //计算s2需要移动的次数c2
    for(i=0;i<s2.size();i++)
    {
        for(j=i+1;j<s2.size();j++)
        {
            if(s2[i]>s2[j])
                c2++;
        }
    }
    //如果两行字符串移动的次数相同,那么按原先固有的位置排序
    return c1!=c2?c1<c2:c1<c2;
}
int main(int argc, char * argv[])
{
    ifstream cin("aaa.txt");
    string s;
    vector<string>v;
    int n,a,b;
    cin>>n;
    int i,j,k;
    int p=0;//块数
    for(i=0;i<n;i++)//读入n组case
    {
        cin.clear();
        cin>>a>>b;
        v.clear();//清空向量
        p++;//增加一块
        for(j=0;j<b;j++)//读入一组中的a行字符串
        {
            cin>>s;
            v.push_back(s);
        }
        //按comp排序规则排序
        sort(v.begin(),v.end(),comp);
        if(p!=1)cout<<endl;//不是第一行,产生一新空行
        //输出v向量中的每个元素
```

```
            for(k=0;k<v.size();k++)
            {
                cout<<v[k]<<endl;
            }
        }
        return 0;
    }
```

4.17.7 汉语翻译

1. 题目

DNA 排 序

在一个系列中,"没排序"的一个尺度是无序对的总数。例如,在字母序列"DAABEC"中,"没排序"的尺度是5,因为 D 比它右边的 4 个字母大,而 E 比它右边的 1 个字母大。这个尺度被称为序列的倒位数量。序列"AACEDGG"只有一个倒位(E 和 D)——即它几乎是有序的——然而,序列"ZWQM"则有 6 个倒位(其序正好完全相反)。

你负责编排 DNA 字符串的序列(系列中只包含 A,C,G 和 T 四个字符)。然而,不是根据字母表顺序来编排,而是根据"没排序"程序即倒位数量来编排,从"排序最好"到"排序最不好"的顺序来编排。所有的字符串长度相同。

问题包含多个测试案例!

多个测试案例前是一个正整数 N,然后是一行空行,接下去是 N 个输入块。每个输入块的格式在问题中描述了。输入块之间用空行隔开。

2. 输入描述

每个输入块的第一行包含两个整数:正整数 n($0 < n \leqslant 50$),表示字符串的长度;另一个正整数是 m($1 < m \leqslant 100$),表示字符串的数目。下面就是 m 行,每行是一个长为 n 的字符串。

3. 输出描述

输出输入数据中的字符串,按从"排序最好"到"排序最不好"的顺序排列。如果两个字符串的倒位数量相同,那么按它们在输入文件中的先后顺序输出。

4. 输入样例

```
1

10 6
AACATGAAGG
TTTTGGCCAA
TTTGGCCAAA
GATCAGATTT
CCCGGGGGGA
ATCGATGCAT
```

5. 输出样例

```
CCCGGGGGA
AACATGAAGG
GATCAGATTT
ATCGATGCAT
TTTTGGCCAA
TTTGGCCAAA
```

4.18 Daffodil Number

4.18.1 链接地址

http://www.realoj.com/网上第 102 题

4.18.2 时空限制

Time Limit: 1000 ms   Resident Memory Limit: 1024 KB   Output Limit: 1024 B

4.18.3 题目内容

The daffodil number is one of the famous interesting numbers in the mathematical world. A daffodil number is a three-digit number whose value is equal to the sum of cubes of each digit. For example. 153 is a daffodil as $153 = 1^3 + 5^3 + 3^3$.

Input

There are several test cases in the input, each case contains a three-digit number.

Output

One line for each case. if the given number is a daffodil number, then output "Yes", otherwise "No".

Sample Input

```
153
610
```

Sample Output

```
Yes
No
```

4.18.4 题目来源

Zhejiang Provincial Programming Contest 2006, Preliminary (Author: LIU, Yaoting)

4.18.5 解题指导

本题是判断一个三位数是否是水仙花数，即它是否等于它各位的立方相加。考查的要点应当是如何将一个数的各位分离出来。通过取余，可以直接将一个数的各位分离。

本题已明确指出，输入数据都是三位数，如果没有指出这点，那么，我们还需要事先判断输入的数据是否是三位数。如果忽略了这点，而数据中存在非三位数的情况，则很难解答正确。所以，看似简单的问题，若不好好考虑一下，也不容易做对。ACM 程序设计就是这样，输入数据我们是没法看得到的，只能根据题意多作精确推理。

4.18.6 参考答案

```
#include <fstream>
#include <iostream>
using namespace std;
int main(int argc, char * argv[])
{
    ifstream cin("aaa.txt");
    int n;
    int a,b,c;
    while(cin>>n)
    {
        a=n/100;
        b=(n-a*100)/10;
        c=n%10;
        if(n==a*a*a + b*b*b + c*c*c)
        {
            cout<<"Yes"<<endl;
        }
        else
        {
            cout<<"No"<<endl;
        }
    }
    return 0;
}
```

4.18.7 汉语翻译

1. 题目

水 仙 花 数

水仙花数是数学界最有名最令人感兴趣的问题之一。如果一个三位数的各位的立方相加等于它本身，那么，这个数就是水仙花数。

例如，153 是一个水仙花数，因为 $153 = 1^3 + 5^3 + 3^3$。

2. 输入描述

输入数据中有多个测试案例，每个测试案例包含一个三位数。

3. 输出描述

一个测试案例占一行。如果给定的数是水仙花数，输出"Yes"，否则输出"No"。

4. 输入样例

   153
   610

5. 输出样例

   Yes
   No

4.19 Error Correction

4.19.1 链接地址

http://www.realoj.com/网上第 103 题

4.19.2 时空限制

Time Limit: 1000 ms   Resident Memory Limit: 1024 KB   Output Limit: 1024 B

4.19.3 题目内容

A boolean matrix has the parity property when each row and each column has an even sum, i.e. contains an even number of bits which are set. Here's a 4×4 matrix which has the parity property:

1 0 1 0
0 0 0 0
1 1 1 1
0 1 0 1

The sums of the rows are 2, 0, 4 and 2. The sums of the columns are 2, 2, 2 and 2.

Your job is to write a program that reads in a matrix and checks if it has the parity property. If not, your program should check if the parity property can be established by changing only one bit. If this is not possible either, the matrix should be classified as corrupt.

Input

The input will contain one or more test cases. The first line of each test case contains one integer n ($n < 100$), representing the size of the matrix. On the next n lines, there will be n integers per line. No other integers than 0 and 1 will occur in the matrix. Input will be terminated by a value of 0 for n.

Output

For each matrix in the input file, print one line. If the matrix already has the parity property, print "OK". If the parity property can be established by changing one bit, print "Change bit (i,j)" where i is the row and j the column of the bit to be changed. Otherwise, print "Corrupt".

Sample Input

```
4
1 0 1 0
0 0 0 0
1 1 1 1
0 1 0 1
4
1 0 1 0
0 0 1 0
1 1 1 1
0 1 0 1
4
1 0 1 0
0 1 1 0
1 1 1 1
0 1 0 1
0
```

Sample Output

```
OK
Change bit (2,3)
Corrupt
```

4.19.4 题目来源

University of Ulm Local Contest 1998

4.19.5 解题思路

本题是判断一个矩阵是否具有奇偶性。这点好办，计算一下每行每列的和是否是偶数就好了，但还需要判断一个没有奇偶性的矩阵是否可以通过修改唯一一位来使之具有奇偶性，这是难点。

当只有一行的和是奇数且只有一列的和为奇数，那么，将这两行交点处的元素修改一下，就变成具有 parity property（奇偶性）的矩阵了。这里还要清楚的一点是，在一个矩阵中，任何一行与任何一列都有且仅有一个交点。

需要指出的是，如果定义一个过大的数组或容器，应当把它们定义为全局变量为宜，这样，可以获得尽可能大的内存分配。这种情况，不宜定义在函数范围内的变量（局部变量）。

4.19.6 参考答案

```cpp
#include <fstream>
#include <iostream>
using namespace std;
int matrix[100][100]; //全局数组
int SL[100];//数组每行的和
int SC[100];//数组每列的和
int main(int argc, char * argv[])
```

```
{
    ifstream cin("aaa.txt");
    //先定义好各种循环变量
    int i,j,PL,PC,CountL,CountC;
    int n;//矩阵行数或列数
    while(cin>>n)
    {
        if(n==0)break;//读入数据结束
        //初始化各种变量和数组
        PL=0;
        PC=0;
        CountL=0;
        CountC=0;
        for(i=0;i<n;i++)
        {
            SL[i]=0;
            SC[i]=0;
        }
        //读入所有数据，同时计算出每行、列的元素和
        for(i=0;i<n;i++)
        {
            for(j=0;j<n;j++)
            {
                cin>>matrix[i][j];
                SL[i]=SL[i]+matrix[i][j];
                SC[j]=SC[j]+matrix[i][j];
            }
        }
        //判断矩阵每行、列的和是否是偶数
        for(i=0;i<n;i++)
        {
            //判断每行的和
            if(SL[i]%2!=0)//和是奇数
            {
                PL=i;//记下行数
                CountL++;//记下次数
            }
            //判断每列的和
            if(SC[i]%2!=0)//和是奇数
            {
                PC=i;//记下列数
                CountC++;//记下次数
            }
        }
        //输出结果
        if(CountL==0 && CountC==0)
            cout<<"OK"<<endl;
        else if(CountL==1 && CountC==1)
            cout<<"Change bit ("<<PL+1<<","<<PC+1<<")"<<endl;
```

```
        else
            cout<<"Corrupt"<<endl;
    }
    return 0;
}
```

4.19.7 汉语翻译

1. 题目

错 误 纠 正

一个布尔矩阵有一种奇偶性,即该矩阵所有行和所有列的和都是一个偶数。下面这个 4×4 的矩阵就具有奇偶性:

1 0 1 0
0 0 0 0
1 1 1 1
0 1 0 1

它所有行的和是 2,0,4 和 2。它所有列的和是 2,2,2 和 2。

你的工作就是编写一个程序,读入这个矩阵并检查它是否具有奇偶性。如果没有,你的程序应当再检查一下它是否可以通过修改一位(把 0 修改为 1,把 1 修改为 0)来使它具有奇偶性。如果不可能,这个矩阵就被认为是破坏了。

2. 输入描述

输入包含多个测试案例。每个测试案例的第一行是一个整数 n($n < 100$),代表该矩阵的大小。在接下来的 n 行中,每行有 n 个整数。矩阵是由 0 或 1 构成的。n 是 0 时,表示输入的结束。

3. 输出描述

对于输入文件中的每个矩阵,打印一行。如果这个矩阵具有奇偶性,那么打印"OK"。如果奇偶性能通过只修改该矩阵中的一位来建立,那么打印"Change bit (i,j)",这里 i 和 j 是被修改的这位的行号和列号。否则,打印"Corrupt"。

4. 输入样例

```
4
1 0 1 0
0 0 0 0
1 1 1 1
0 1 0 1
4
1 0 1 0
0 0 1 0
1 1 1 1
0 1 0 1
4
1 0 1 0
```

```
0 1 1 0
1 1 1 1
0 1 0 1
0
```

5. 输出样例

```
OK
Change bit (2,3)
Corrupt
```

4.20 Martian Addition

4.20.1 链接地址

http://www.realoj.com/网上第 104 题

4.20.2 时空限制

Time Limit: 1000 ms   Resident Memory Limit: 1024 KB   Output Limit: 1024 B

4.20.3 题目内容

In the 22nd Century, scientists have discovered intelligent residents live on the Mars. Martians are very fond of mathematics. Every year, they would hold an Arithmetic Contest on Mars (ACM). The task of the contest is to calculate the sum of two 100-digit numbers, and the winner is the one who uses least time. This year they also invite people on Earth to join the contest.

As the only delegate of Earth, you're sent to Mars to demonstrate the power of mankind. Fortunately you have taken your laptop computer with you which can help you do the job quickly. Now the remaining problem is only to write a short program to calculate the sum of 2 given numbers. However, before you begin to program, you remember that the Martians use a 20-based number system as they usually have 20 fingers.

Input

You're given several pairs of Martian numbers, each number on a line. Martian number consists of digits from 0 to 9, and lower case letters from a to j (lower case letters starting from a to present 10, 11, ..., 19). The length of the given number is never greater than 100.

Output

For each pair of numbers, write the sum of the 2 numbers in a single line.

Sample Input

```
1234567890
abcdefghij
99999jjjjj
9999900001
```

Sample Output

```
bdfi02467j
iiiij00000
```

4.20.4 题目来源

Zhejiang University Local Contest 2002, Preliminary

4.20.5 解题思路

本题要求进行 100 位以内的二十进制数字相加运算。加数和被加数的长度不一定相等。这是超长数字的相加。超长计算我们通常把它们当做字符串来处理，因为字符串（string）并没有长度的限制，可以处理任意位数的加法运算。

先把加数和被加数这两个字符串反转，再从第 0 位开始到末尾，两位依次相加，每两位计算的结果放到一个向量中，如果两位相加的结果超过 19，那么需要向下一位产生进位。全部位计算完成后，再反向输入字符串即可。

按这样的思路来编写程序，要有耐心，只要头脑清楚，编程的基本功扎实，按照普通加法的思路，一定能做出来。超长数字的减法、乘法和除法都可依此方式来计算。

4.20.6 参考答案

```cpp
#include <fstream>
#include <iostream>
#include <vector>//向量
#include <string>//字符容器
#include <algorithm>//算法
using namespace std;
int main(int argc, char * argv[])
{
    ifstream cin("aaa.txt");
    string sa,sb,t;
    vector<int>v;
    int i;
    int a,b,sum;
    int flag;//进位标记
    while(cin>>sa>>sb)
    {
        //初始化各种变量和容器
        flag=0;//没有进位
        v.clear();//清空结果向量
        sum=0;//两位数的和
        //反转两个字符容器
        reverse(sa.begin(),sa.end());
        reverse(sb.begin(),sb.end());
        //sa 里放长串，sb 里放短串
```

```
if(sa.size()<sb.size())
{
    t=sa;sa=sb;sb=t;
}
//从个位起，一位一位相加
for(i=0;i<sa.size();i++)
{
    //0 1 2 3 4 5 6 7 8 9 10 11 12 13 14 15 16 17 18 19
    //0 1 2 3 4 5 6 7 8 9 a  b  c  d  e  f  g  h  i  j
    if     (sa[i]=='0')a=0;
    else if(sa[i]=='1')a=1;
    else if(sa[i]=='2')a=2;
    else if(sa[i]=='3')a=3;
    else if(sa[i]=='4')a=4;
    else if(sa[i]=='5')a=5;
    else if(sa[i]=='6')a=6;
    else if(sa[i]=='7')a=7;
    else if(sa[i]=='8')a=8;
    else if(sa[i]=='9')a=9;
    else if(sa[i]=='a')a=10;
    else if(sa[i]=='b')a=11;
    else if(sa[i]=='c')a=12;
    else if(sa[i]=='d')a=13;
    else if(sa[i]=='e')a=14;
    else if(sa[i]=='f')a=15;
    else if(sa[i]=='g')a=16;
    else if(sa[i]=='h')a=17;
    else if(sa[i]=='i')a=18;
    else if(sa[i]=='j')a=19;

    if(i>=sb.size())b=0;//已超过b串的最长一位
    else
    {
        if     (sb[i]=='0')b=0;
        else if(sb[i]=='1')b=1;
        else if(sb[i]=='2')b=2;
        else if(sb[i]=='3')b=3;
        else if(sb[i]=='4')b=4;
        else if(sb[i]=='5')b=5;
        else if(sb[i]=='6')b=6;
        else if(sb[i]=='7')b=7;
        else if(sb[i]=='8')b=8;
        else if(sb[i]=='9')b=9;
        else if(sb[i]=='a')b=10;
        else if(sb[i]=='b')b=11;
        else if(sb[i]=='c')b=12;
        else if(sb[i]=='d')b=13;
        else if(sb[i]=='e')b=14;
        else if(sb[i]=='f')b=15;
        else if(sb[i]=='g')b=16;
```

```
            else if(sb[i]=='h')b=17;
            else if(sb[i]=='i')b=18;
            else if(sb[i]=='j')b=19;
        }
        sum=a+b+flag;//求两位的和
        if(sum>19)
        {
            flag=1;//进位
            sum=sum-20;//保留位
        }
        else
        {
            flag=0;
        }
        v.push_back(sum);//保存保留位
    }
    if(flag==1)v.push_back(1);
    //输出结果
    for(i=v.size()-1;i>=0;i--)
    {
        if(v[i]<10)cout<<v[i];
        else if(v[i]==10)cout<<"a";
        else if(v[i]==11)cout<<"b";
        else if(v[i]==12)cout<<"c";
        else if(v[i]==13)cout<<"d";
        else if(v[i]==14)cout<<"e";
        else if(v[i]==15)cout<<"f";
        else if(v[i]==16)cout<<"g";
        else if(v[i]==17)cout<<"h";
        else if(v[i]==18)cout<<"i";
        else if(v[i]==19)cout<<"j";
    }
    cout<<endl;
    }
    return 0;
}
```

4.20.7 汉语翻译

1. 题目

火 星 加 法

在 22 世纪，科学家发现，火星上有智慧的居民。火星人都非常喜欢数学。每年，他们都要在火星上举行一次算术竞赛（ACM）。竞赛的内容是计算两个 100 位数字的和，该大奖的得主是使用时间最少的人。今年他们还请地球上的人参加竞赛。

作为地球上唯一的代表，您被发送到火星上，以展示人类的力量。幸好你带上了你的笔记本电脑，它可以帮助你做得很快。现在剩下的问题是，你只需编写一个简短的程序来计算两个给定的数字的总和。不过，在开始编写程序前，要记住，火星人使用的是二十进制的数字系统，因为他们通常有 20 根手指。

2. 输入描述

给你若干对火星数字，每个数字占一行。

火星数字由 0 到 9，以及从 a 到 j 的小写字母组成（小写字母 a 到 j 分别代表 10，11，…，19）。

给定的数字的长度不会超过 100 位。

3. 输出描述

对于每对数字，在一行上输出它们的和。

4. 输入样例

```
1234567890
abcdefghij
99999jjjjj
9999900001
```

5. 输出样例

```
bdfi02467j
iiiij00000
```

4.21 FatMouse' Trade

4.21.1 链接地址

http://www.realoj.com/网上第 105 题

4.21.2 时空限制

Time Limit: 1000 ms   Resident Memory Limit: 1024 KB   Output Limit: 1024 B

4.21.3 题目内容

FatMouse prepared M pounds of cat food, ready to trade with the cats guarding the warehouse containing his favorite food, JavaBean.

The warehouse has N rooms. The i-th room contains $J[i]$ pounds of JavaBeans and requires $F[i]$ pounds of cat food. FatMouse does not have to trade for all the JavaBeans in the room, instead, he may get $J[i]*a\%$ pounds of JavaBeans if he pays $F[i]*a\%$ pounds of cat food. Here a is a real number. Now he is assigning this homework to you: tell him the maximum amount of JavaBeans he can obtain.

Input

The input consists of multiple test cases. Each test case begins with a line containing two non-negative integers M and N. Then N lines follow, each contains two non-negative integers $J[i]$ and $F[i]$ respectively. The last test case is followed by two −1's. All integers are not greater than 1000.

Output

For each test case, print in a single line a real number accurate up to 3 decimal places, which is the maximum amount of JavaBeans that FatMouse can obtain.

Sample Input

```
5 3
7 2
4 3
5 2
20 3
25 18
24 15
15 10
-1 -1
```

Sample Output

```
13.333
31.500
```

4.21.4　题目来源

Zhejiang Provincial Programming Contest 2004 (Author: CHEN, Yue)

4.21.5　解题指导

本题要求输出最大的交易量，并保留 3 位小数。这样，我们使用 $J[i]$ 除以 $F[i]$ 就得到了 a，那么，交易的时候，为了获得最多的 JavaBean，那么，要先交易 a 大的，这样就确保了能交易到最多的 JavaBean。

把数据读入结构体中，再将结构体作为向量的元素，再按 a 由大到小的顺序给向量排序，然后依次进行计算，这种方法是比较容易的。

4.21.6　参考答案

```cpp
#include <fstream>
#include <iostream>
#include <vector>
#include <algorithm>
using namespace std;
//数据存取结构
struct Mouse{
    double J;
    double F;
    double a;
};
```

```cpp
//自定义排序比较规则
bool Comp(const Mouse &d1,const Mouse &d2)
{
    //按 a 由大到小排序
    if(d1.a!=d2.a)return d1.a>d2.a;
    //按 F 由大到小排序
    else return d1.F<d2.F;
}
int main(int argc, char * argv[])
{
    ifstream cin("aaa.txt");
    vector<Mouse> v;
    Mouse mouse;
    int m,n,i;
    //三位输出精度
    cout.precision(3);
    double sum;
    while(cin>>m>>n)
    {
        if(m==-1 && n==-1)break;//读取结束
        v.clear();
        sum=0.0;
        //读入数据到向量中
        for(i=0;i<n;i++)
        {
            cin>>mouse.J>>mouse.F;
            mouse.a=mouse.J/mouse.F;
            v.push_back(mouse);//往向量里增加一元素
        }
        //使用自定义排序规则 Comp 给向量排序
        sort(v.begin(),v.end(),Comp);
        //计算交换数量
        for(i=0;i<v.size();i++)
        {
            if(m>=v[i].F)
            {
                sum=sum + v[i].J;
                m=m - v[i].F;
            }
            else
            {
                sum=sum + m * v[i].a;
                break;
            }
        }
        //定点三位精度输出数据
        cout<<fixed<<sum<<endl;
    }
    return 0;
}
```

4.21.7 汉语翻译

1. 题目

FatMouse 的交易

FatMouse 准备了 M 磅猫食，准备与守卫仓库的猫们进行交易，仓库里有他最爱吃的食物 JavaBean。

仓库有 N 间房间。第 i 间房间里有 $J[i]$ 磅 JavaBean 且需要 $F[i]$ 磅猫食来交易。FatMouse 不必把每个房间里的 JavaBean 全部用于交易，相反，它可以付给猫 $F[i]*a\%$ 磅猫食，从而换得 $J[i]*a\%$ 磅 JavaBean。其中，a 是一个实数。现在它给你布置一个家庭作业：请你告诉它最多能够获得多少磅 JavaBean。

2. 输入描述

输入包含多个测试案例。每个测试案例的开头一行是两个非负整数 M 和 N。接下去的 N 行中，每行包含两个非负整数 $J[i]$ 和 $F[i]$。最后一个测试案例是两个 -1。所有的整数的值不会超过 1000。

3. 输出描述

对于每个测试案例，在一行上打印出一个 3 位小数的实数，这个实数是 FatMouse 能够交易到的最大数量的 JavaBean。

4. 输入样例

```
5 3
7 2
4 3
5 2
20 3
25 18
24 15
15 10
-1 -1
```

5. 输出样例

```
13.333
31.500
```

4.22 List the Books

4.22.1 链接地址

http://www.realoj.com/网上第 106 题

4.22.2 时空限制

Time Limit: 1000 ms   Resident Memory Limit: 1024 KB   Output Limit: 1024 B

4.22.3 题目内容

Jim is fond of reading books, and he has so many books that sometimes it's hard for him to manage them. So he is asking for your help to solve this problem.

Only interest in the name, press year and price of the book, Jim wants to get a sorted list of his books, according to the sorting criteria.

Input

The problem consists of multiple test cases.

In the first line of each test case, there's an integer n that specifies the number of books Jim has. n will be a positive integer less than 100. The next n lines give the information of the books in the format *Name Year Price*. *Name* will be a string consisting of at most 80 characters from alphabet, *Year* and *Price* will be positive integers. Then comes the sorting criteria, which could be *Name*, *Year* or *Price*.

Your task is to give out the book list in ascending order according to the sorting criteria in non-ascendent order.

Note: That *Name* is the first criteria, *Year* is the second, and *Price* the third. It means that if the sorting criteria is *Year* and you got two books with the same *Year*, you'd sort them according to their *Name*. If they equals again, according to their *Price*. No two books will be same in all the three parameters.

Input will be terminated by a case with $n = 0$.

Output

For each test case, output the book list, each book in a line. In each line you should output in the format *Name Year Price*, the three parameters should be seperated by just **ONE** space.

You should output a blank line between two test cases.

Sample Input

```
3
LearningGNUEmacs 2003 68
TheC++StandardLibrary 2002 108
ArtificialIntelligence 2005 75
Year
4
GhostStory 2001 1
WuXiaStory 2000 2
SFStory 1999 10
WeekEnd 1998 5
Price
0
```

Sample Output

```
TheC++StandardLibrary 2002 108
LearningGNUEmacs 2003 68
ArtificialIntelligence 2005 75

GhostStory 2001 1
WuXiaStory 2000 2
WeekEnd 1998 5
SFStory 1999 10
```

4.22.4 题目来源

Zhejiang University Local Contest 2006, Preliminary (Author: DAI, Wenbin)

4.22.5 解题指导

本题是三维排序问题。Name，Year 和 Price 分别是默认的第一、第二和第三排序。但本例又很特殊，第一排序是由输入指定的，是 Name，Year 和 Price 三者之一。比如，如果 Name 是排序标准，那么，三个排序顺序是 Name，Year 和 Price；当 Year 是排序标准时，三个排序顺序是 Year，Name 和 Price；当 Price 是排序标准时，三个排序顺序是 Price，Name 和 Year。

对于这三种排序标准，我们编写三个比较函数，这样就显得比较灵活，程序实现起来也比较容易。

而书的信息，则写到一个结构体变量中，而结构体变量又作为向量的元素，最后，只要根据排序标准，在排序算法中选择相应的比较函数，就完成了书的排序了。

本题对泛型编程的优势发挥得淋漓尽致，如果不采用泛型编程，则会变得相当困难。

4.22.6 参考答案

```cpp
#include <fstream>
#include <iostream>
#include <vector>
#include <string>
#include <algorithm>
using namespace std;
//数据存取结构
struct Book{
    string Name;
    int Year;
    int Price;
};
//自定义比较函数,Name 主序
bool CompName(const Book &b1,const Book &b2)
{
    if(b1.Name!=b2.Name)return b1.Name<b2.Name;
    else if(b1.Year!=b2.Year)return b1.Year<b2.Year;
```

```cpp
        else return b1.Price<b2.Price;
}
//自定义比较函数，Year 主序
bool CompYear(const Book &b1,const Book &b2)
{
    if(b1.Year!=b2.Year)return b1.Year<b2.Year;
    else if(b1.Name!=b2.Name)return b1.Name<b2.Name;
    else return b1.Year<b2.Year;
}
//自定义比较函数，Price 主序
bool CompPrice(const Book &b1,const Book &b2)
{
    if(b1.Price!=b2.Price)return b1.Price<b2.Price;
    else if(b1.Name!=b2.Name)return b1.Name<b2.Name;
    else return b1.Year<b2.Year;
}
int main(int argc, char * argv[])
{
    ifstream cin("aaa.txt");
    vector<Book> v;
    Book book;
    string sorting;//第一排序
    int n;
    int i;
    int line=0;
    while(cin>>n)
    {
        if(n==0)break;
        line++;//案例号
        v.clear();//清空向量
        //读入一个案例数据到向量中
        for(i=0;i<n;i++)
        {
            cin>>book.Name>>book.Year>>book.Price;
            v.push_back(book);
        }
        //读入排序标准
        cin>>sorting;
        //给向量排序,要据主排序确定比较函数
        if(sorting=="Name")
            sort(v.begin(),v.end(),CompName);
        else if(sorting=="Year")
            sort(v.begin(),v.end(),CompYear);
        else if(sorting=="Price")
            sort(v.begin(),v.end(),CompPrice);
        //输出结果
        if(line!=1)cout<<endl;//不是第一行，先输出一空行
        for(i=0;i<v.size();i++)
        {
            cout<<v[i].Name<<" "<<v[i].Year<<" "<<v[i].Price<<endl;
```

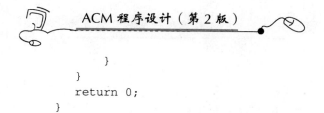

```
            }
        }
        return 0;
    }
```

4.22.7 汉语翻译

1. 题目

列　书

Jim 很喜欢读书，他有很多书，有时他很难管理它们。因此，他请求你的帮助来解决这个问题。

Jim 想以书名、出版年份和价格为关键词，按照排序标准把书进行排序。

2. 输入描述

本题包含多个测试案例。

在每个测试案例的第一行是一个整数 n，表示 Jim 有多少本书。n 是一个小于 100 的正整数。接下去的 n 行以 Name Year Price 的格式给出了书的信息。Name 是一个只包含字母且长度至多为 80 的字符串，Year 和 Price 是正整数。最后一项是排序标准，可能是 Name，Year 或 Price。

你的任务就是以排序标准给书按升序排列。

注意：Name 是第一排序标准，Year 是第二排序标准，Price 是第三排序标准。如果排序标准是 Year，且两本书的 Year 相同，那么，你就应该给它们按 Name 排序。如果 Name 还是相同，那么，就应该按 Price 排序。不存在两本书的三个参数都相同的情况。

当 n =0 时，表示输入的结束。

3. 输出描述

对于每个测试案例，输出书的列表，每本书打印一行。每行你应该按 Name Year Price 的格式打印，这参数间应该以一个空格隔开。

在两个测试案例之间要有一个空行。

4. 输入样例

```
3
LearningGNUEmacs 2003 68
TheC++StandardLibrary 2002 108
ArtificialIntelligence 2005 75
Year
4
GhostStory 2001 1
WuXiaStory 2000 2
SFStory 1999 10
WeekEnd 1998 5
Price
0
```

5. 输出样例

```
TheC++StandardLibrary 2002 108
LearningGNUEmacs 2003 68
ArtificialIntelligence 2005 75

GhostStory 2001 1
WuXiaStory 2000 2
WeekEnd 1998 5
SFStory 1999 10
```

4.23 Head-to-Head Match

4.23.1 链接地址

http://www.realoj.com/网上第 107 题

4.23.2 时空限制

Time Limit: 1000 ms   Resident Memory Limit: 1024 KB   Output Limit: 1024 B

4.23.3 题目内容

Our school is planning to hold a new exciting computer programming contest. During each round of the contest, the competitors will be paired, and compete head-to-head. The loser will be eliminated, and the winner will advance to next round. It proceeds until there is only one competitor left, who is the champion. In a certain round, if the number of the remaining competitors is not even, one of them will be chosed randomly to advance to next round automatically, and then the others will be paired and fight as usual. The contest committee want to know how many rounds is needed to produce to champion, then they could prepare enough problems for the contest.

Input

The input consists of several test cases. Each case consists of a single line containing a integer N—the number of the competitors in total. $1 \leqslant N \leqslant 2\ 147\ 483\ 647$. An input with 0 (zero) signals the end of the input, which should not be processed.

Output

For each test case, output the number of rounds needed in the contest, on a single line.

Sample Input

```
8
16
15
0
```

Sample Output

```
3
4
4
```

4.23.4 题目来源

Zhejiang University Local Contest 2006, Preliminary (Author: YANG, Chao)

4.23.5 解题指导

两两比赛，胜出一人。胜者再晋级参加下一轮比赛。如果该轮总人数是奇数，则随机选出一人直接参加下一轮比赛。求比赛所需要的总轮数。

本题如果算法不好，容易出现超时错误。对于 N 个参赛者，两两比赛，需要 n 轮比赛（当 $2^n \geqslant N$ 时）。

这里特别注意，N 的取值范围是 $1 \leqslant N \leqslant 2\ 147\ 483\ 647$，也就是说，可以是 1 人参赛。如果是 1 人，比赛的轮数是 0 轮，即不用举行比赛。

4.23.6 参考答案

```cpp
#include <fstream>
#include <iostream>
#include <cmath>
using namespace std;
int main(int argc, char * argv[])
{
    ifstream cin("aaa.txt");
    //pow乘方函数的参数需要定义为double 类型
    double i,n,count;
    while(cin>>n)
    {
        if(n==0)break;
        count=0;
        for(i=0;i<32;i++)
        {
            if(pow(2,i)>=n)
            {
                count=i;
                break;
            }
        }
        cout<<count<<endl;
    }
    return 0;
}
```

4.23.7 汉语翻译

1. 题目

两 两 比 赛

我们学校正计划举行一场全新的激动人心的计算机程序设计竞赛。在每一轮比赛中，参赛者都是成对的，两两比赛。输者将被淘汰，赢者将自动晋级到下一轮比赛中。比赛一直进行到只剩下一个人为止，这个人就是冠军。在一轮比赛中，如果比赛人数不是偶数，那么将随机选择一个参赛者自动晋级到下一轮比赛中，而其他人则还是一对一地完成本轮比赛。竞赛委员会想知道要产生冠军需要进行多少轮比赛，以便可以准备好足够的竞赛题目。

2. 输入描述

输入包含多个测试案例。每个测试案例在一行上，是一个整数 N，N 就是参赛总人数。$1 \leqslant N \leqslant 2\,147\,483\,647$。输入 0 表示输入的结束，不需要处理。

3. 输出描述

对于每个测试案例，在一行上输出比赛需要的总轮数。

4. 输入样例

```
8
16
15
0
```

5. 输出样例

```
3
4
4
```

4.24 Windows Message Queue

4.24.1 链接地址

http://www.realoj.com/网上第 108 题

4.24.2 时空限制

Time Limit: 1000 ms   Resident Memory Limit: 1024 KB   Output Limit: 1024 B

4.24.3 题目内容

Message queue is the basic fundamental of windows system. For each process, the system maintains a message queue. If something happens to this process, such as mouse click, text change, the system will add a message to the queue. Meanwhile, the process will do a loop for getting message from the queue according to the priority value if it is not empty. Note that the less priority value means the higher priority. In this problem, you are asked to simulate the message queue for putting messages to and getting message from the message queue.

Input

There's only one test case in the input. Each line is a command, "GET" or "PUT", which means getting message or putting message. If the command is "PUT", there're one string means the message name and two integer means the parameter and priority followed by. There will be at most 60 000 command. Note that one message can appear twice or more and if two messages have the same priority, the one comes first will be processed first. (i.e., FIFO for the same priority.) Process to the end-of-file.

Output

For each "GET" command, output the command getting from the message queue with the name and parameter in one line. If there's no message in the queue, output "EMPTY QUEUE!". There's no output for "PUT" command.

Sample Input

```
GET
PUT msg1 10 5
PUT msg2 10 4
GET
GET
GET
```

Sample Output

```
EMPTY QUEUE!
msg2 10
msg1 10
EMPTY QUEUE!
```

4.24.4 题目来源

Zhejiang University Local Contest 2006, Preliminary (Author: ZHOU, Ran)

4.24.5 解题指导

本题是模拟 Windows 处理消息队列,很有意义。解题中主要遇到的问题是超时错误。根据本题的特点,宜使用优先队列容器来实现。

另外，光使用优先队列容器还不行，必须采用 scanf 和 printf 输入输出，否则还会超时，因为本题的数据量大，多达 60 000 行字符串需要处理，需要输入输出。

4.24.6 参考答案

```cpp
#include <iostream>
#include <queue>
using namespace std;
struct Message{
    char Name[100];
    int Data;
    int Priority;
    //重载<操作符，自定义排序规则
    bool operator < (const Message &a) const
    {
        return a.Priority<Priority;
    }
};
priority_queue<Message> v;
int main(int argc, char * argv[])
{
    char command[100];
    Message message;
    while(scanf("%s",command)!=EOF)
    {
        if(strcmp(command,"GET")==0)
        {
            //队列为空
            if(v.size()==0)
                printf("EMPTY QUEUE!\n");
                //使用下面这句话不会超时
                //cout<<"EMPTY QUEUE!"<<endl;
            else
            {
                //cout<<<" "<<<<endl;
                printf("%s %d\n",v.top().Name,v.top().Data);
                //使用下面这句话会超时
                //cout<<v.top().Name<<" "<<v.top().Data<<endl;
                //出队列操作，即将当前消息清除
                v.pop();
            }
        }
        else if(strcmp(command,"PUT")==0)
        {
            scanf("%s%d%d",&message.Name,&message.Data,&message.Priority);
            v.push(message);
        }
    }
    return 0;
}
```

4.24.7 汉语翻译

1. 题目

Windows 消息队列

消息队列是 Windows 操作系统的基石。对于每个进程，系统维持了一个消息队列。如果一个进程发生了某一事件，如单击鼠标，文本改变，系统将增加一个消息到队列里。同时，如果队列里不是空的，那么，进程将一直循环地从队列里按优先值抓取消息。注意，数值小意味着高的优先级。在本题中，要求你模拟消息队列，把消息放到消息队列中，或从消息队列里抓取消息。

2. 输入描述

只有一个测试案例。每行是一条命令，"GET"或"PUT"表示从消息队列里抓取消息或把消息放入消息队列。如果是"PUT"命令，后面跟着一个字符串（表示消息名称）和两个整数（分别表示消息的参数和该消息的优先级）。案例中的命令最多达 60 000 条。注意，一条命令可以重复出现多次，如果两条消息的优先级相同，先进入消息队列的那条先被处理。（优先级相同的命令，是先进先出的。）一直处理到文件结尾。

3. 输出描述

对于每条"GET"命令，直接输出它抓取的消息的名称和参数在一行上。如果消息队列是空的，那么直接输出"EMPTY QUEUE!"。对于"PUT"命令，不需要输出什么。

4. 输入样例

```
GET
PUT msg1 10 5
PUT msg2 10 4
GET
GET
GET
```

5. 输出样例

```
EMPTY QUEUE!
msg2 10
msg1 10
EMPTY QUEUE!
```

4.25 Language of FatMouse

4.25.1 链接地址

http://www.realoj.com/ 网上第 109 题

4.25.2 时空限制

Time Limit: 1000 ms　　Resident Memory Limit: 1024 KB　　Output Limit: 1024 B

4.25.3 题目内容

We all know that FatMouse doesn't speak English. But now he has to be prepared since our nation will join WTO soon. Thanks to Turing we have computers to help him.

Input Specification

Input consists of up to 100 005 dictionary entries, followed by a blank line, followed by a message of up to 100 005 words. Each dictionary entry is a line containing an English word, followed by a space and a FatMouse word. No FatMouse word appears more than once in the dictionary. The message is a sequence of words in the language of FatMouse, one word on each line. Each word in the input is a sequence of at most 10 lowercase letters.

Output Specification

Output is the message translated to English, one word per line. FatMouse words not in the dictionary should be translated as "eh".

Sample Input

```
dog ogday
cat atcay
pig igpay
froot ootfray
loops oopslay

atcay
ittenkay
oopslay
```

Output for Sample Input

```
cat
eh
loops
```

4.25.4 题目来源

Zhejiang University Training Contest 2001

4.25.5 解题思路

本题是词典搜索问题。如果采用传统的数组结构来保存词典，那么，多达 100 005 条词的词典，肯定会产生超时错误。

本题应当采用映照容器来解决。映照容器的键值应当是 FatMouse 的单词，其映照数据应当是该 FatMouse 单词对应的英文单词，这样，映照建立好后，直接在映照容器中查询

FatMouse 的单词就好了。由于映照容器的数据结构是平衡检索二叉树（黑白树），检索速度极快，因而不会发生超时的错误。

4.25.6 参考答案

```cpp
#include <fstream>
#include <iostream>
#include <string>
#include <map>
using namespace std;
//采用gets()输入比getline()快得多,但gets()不能从cin中读取数据,调试上不易
//gets函数用法:
//从stdin流中读取字符串，直至接受到换行符或EOF时停止,
//并将读取的结果存放在str指针所指向的字符数组中。
//换行符不作为读取串的内容，读取的换行符被转换为null值，并由此来结束字符串
int main(int argc, char * argv[])
{
    //ifstream cin("aaa.txt");
    string s;
    char ss[100],s1[100],s2[100];
    int x,y;
    //map 映照容器:键值+映照数据
    map<string,string>m;
    //迭代器
    map<string,string>::iterator p;
    //while(getline(cin,s))
    while(gets(ss))
    {
        s=ss;
        //如果读入一行为空行(只包含一个回车符)则结束循环
        if(s=="")break;
        else
        {
            //sscanf与scanf类似，都是用于输入的，只是后者以屏幕(stdin)为输入源,
            //前者以固定字符串为输入源，默认以空格来分开字符串
            sscanf(s.c_str(),"%s %s",s1,s2);
            m[s2]=s1;
        }
    }
    //while(cin>>s)
    while(gets(ss))
    {
        s=ss;
        //从映照容器中查找键值
        p=m.find(s);
        if(p!=m.end())
            cout<<m[s]<<endl;
        else
            cout<<"eh"<<endl;
```

```
    }
    return 0;
}
```

4.25.7 汉语翻译

1. 题目

FatMouse 的语言

我们都知道 FatMouse 不会说英语。但他现在已经有所准备，因为我们的国家正要加入 WTO 了。感谢图灵（Turing），我们拿计算机来帮助它们。

2. 输入详细描述

输入数据先是多达 100 005 条词典条目，再输入一空行，然后是多达 100 005 个单词。每个字典条目在一行上，包含一个英文单词，后边跟着一个空格和一个 FatMouse 的语言。FatMouse 的单词不会在词典里重复出现。信息则是 FatMouse 语言的一系列单词，一个单词在一行上。输入数据中的每个单词是一串至多 10 个小写字母的系列。

3. 输出详细描述

把信息翻译为英文输出，一个单词一行。FatMouse 的单词如果在词典里找不到，应当输入"eh"。

4. 输入样例

```
dog ogday
cat atcay
pig igpay
froot ootfray
loops oopslay

atcay
ittenkay
oopslay
```

5. 输出样例

```
cat
eh
loops
```

4.26 Palindromes

4.26.1 链接地址

http://www.realoj.com/网上第 110 题

4.26.2 时空限制

Time Limit: 1000 ms   Resident Memory Limit: 1024 KB   Output Limit: 1024 B

4.26.3 题目内容

A regular palindrome is a string of numbers or letters that is the same forward as backward. For example, the string "ABCDEDCBA" is a palindrome because it is the same when the string is read from left to right as when the string is read from right to left.

Now give you a string S, you should count how many palindromes in any consecutive substring of S.

Input

There are several test cases in the input. Each case contains a non-empty string which has no more than 5000 characters.

Proceed to the end of file.

Output

A single line with the number of palindrome substrings for each case.

Sample Input

```
aba
aa
```

Sample Output

```
4
3
```

4.26.4 题目来源

Zhejiang Provincial Programming Contest 2006 (Author: LIU, Yaoting)

4.26.5 解题思路

本题要求统计一个字符串中包含多少个回文子串。首先我们来确定子串的概念：一个字符串的子串，就是指它本身的各个部分。如字符串"aba"的子串有"a"、"b"、"a"、"ab"、"ba"和"aba"。

再来看回文，回文就是从左读到右和从右读到左都是一样的，长度为1的字符串也是回文。如"a"、"s"、"aa"、"aba"和"aabaa"等都是回文。

本题在一个字符串中，单个字符也被认为是回文子串，相同的重复的子串也需要计算在内。下表列出了几种常见的回文子串情况。

几种常见的回文子串

字 符 串	包含的回文子串	回文子串的总数目
a	a	1
ab	a、b	2
aba	a、b、a、aba	4
aaa	a、a、a、aa、aa、aaa	6

  本题要求判断一个字符串中的所有的子串是否是回文子串。如果用常规方法做，肯定会出现超时错误。这里采用由中心向外扩散的方法去判断一个子串是否是回文子串：如果最中心的子串不是回文，那么，立即终止，不必去判断向外围扩散的子串了，这就大大节约了时间。

  下面以"abaa"为例，讲解由中心向外扩散方法，如下图所示。

```
        从左钉到右    │  从右钉到左
        abaa  0-3    │  abaa  2-0
        abaa  1-3    │  abaa  1-0
        abaa  2-3    │
        
        由中心最小的子串向外
        扩散，中心不是回文，
        就不必向外扩散
```

判断是否为回文子串的方法

（1）从左往右，钉住最后一个字符。

"abaa"串：先考查中心子串"ba"不是回文串，就可以判定"abaa"不是回文串；

"baa"串：先考查中心子串"baa"不是回文，它是最外子串，不必向外扩散；

"aa"串：考查中心子中"aa"是回文，它是最外子串，不必向外扩散。

（2）从右边倒数第二个字符往左，钉住第一个字符。

"aba"串：考查中心子串"aba"，是回文，它是最外子串，不必向外扩散；

"ab"串：考查中心子串"ab"，不是回文，它是最外子串，不必向外扩散。

  这样下来，加上单个子串"a"、"b"、"a"、"a" 4个，"abaa"中共包含6个回文子串。

4.26.6 参考答案

```cpp
#include <fstream>
#include <iostream>
using namespace std;
int main(int argc, char * argv[])
{
    ifstream cin("aaa.txt");
    char s[5000];
```

```cpp
int p,i,Half,Left,Right,Count;
while(cin>>s){
    i=strlen(s);
    Count=0;//总回文数
    //从左到右钉住最后一个
    for(p=0;p<i-1;p++){
        Half=((i-1)-p)/2;
        //如果子串是奇数个字符
        if(((i-1)-p)%2==0){
            Left=p+Half-1;
            Right=p+Half+1;
        }
        else{//如果子串是偶数个字符
            Left=p+Half;
            Right=p+Half+1;
        }
        while(Left>=p){
            if(s[Left]==s[Right]){
                Count++;//发现了一个回文串
                Left--;
                Right++;
            }
            else{//如果不相等,立即中止,由中心向外扩散不可能会有回文串
                break;
            }
        }
    }
    //从右到左钉住第一个
    for(p=i-2;p>=1;p--){
        Half=p/2;
        //如果子串是奇数个
        if(p%2==0){
            Left=Half-1;
            Right=Half+1;
        }
        else{//如果子串是偶数个
            Left=Half;
            Right=Half+1;
        }
        while(Left>=0){
            if(s[Left]==s[Right]){
                Count++;//发现了一个回文串
                Left--;
                Right++;
            }
            else{//如果不相等,立即中止,由中心向外扩散不可能会有回文串
                break;
            }
        }
    }
```

```
        printf("%d\n",Count+i);
    }
    return 0;
}
```

4.26.7 汉语翻译

1. 题目

回 文

一个规则回文是一个由数字或字母组成的字符串,将它从左读到右和从右读到左都是一样的。例如,字符串"ABCDEDCBA"就是一个回文,因为从左读到右和从右读到左都是一样的。

现在给你一个字符串 S,你要计算出它有多少个回文子串。

2. 输入描述

输入数据中有多个测试案例。每个案例是一个非空且长度不超过 5 000 的字符串。
处理到文件结尾。

3. 输出描述

在每行上打印出该字符串中回文子串的个数。

4. 输入样例

```
aba
aa
```

5. 输出样例

```
4
3
```

4.27 Root of the Problem

4.27.1 链接地址

http://www.realoj.com/网上第 111 题

4.27.2 时空限制

Time Limit: 1000 ms   Resident Memory Limit: 1024 KB   Output Limit: 1024 B

4.27.3 题目内容

Given positive integers B and N, find an integer A such that A^N is as close as possible to B. (The result A is an approximation to the Nth root of B.) Note that A^N may be less than, equal to, or greater than B.

Input

The input consists of one or more pairs of values for B and N. Each pair appears on a single line, delimited by a single space. A line specifying the value zero for both B and N marks the end of the input. The value of B will be in the range 1 to 1,000,000 (inclusive), and the value of N will be in the range 1 to 9 (inclusive).

Output

For each pair B and N in the input, output A as defined above on a line by itself.

Example Input	Example Output
4 3	1
5 3	2
27 3	3
750 5	4
1000 5	4
2000 5	4
3000 5	5
1000000 5	16
0 0	

4.27.4 题目来源

Problem Source: Mid-Central USA 2006

4.27.5 解题思路

本题要求满足 $A^N = B$ 式，且最逼近 B 的 A 值。比较简单，细节方面注意一下就可以了。

$A^N = B$，那么，$A = \sqrt[N]{B}$，或者 $A = B^{\frac{1}{N}}$，这样，就可以采用 cmath 中的 pow 幂函数：double pow(double,double)。

有一点需要特别注意，如果一个 double 类型自动转换为 int 类型，那么，小数部分不会向整数部分四舍五入，而是采用截取整数的方法。

4.27.6 参考答案

```
#include <fstream>
#include <iostream>
#include <cmath>
using namespace std;
int main(int argc, char * argv[])
{
    ifstream cin("aaa.txt");
    double B,N;
    int A;
    while(cin>>B>>N)
    {
        if(B==0 && N==0)break;
```

```
            A=pow(B,1/N);
            if(B-pow(A,N)<pow(A+1,N)-B)
                cout<<A<<endl;
            else
                cout<<A+1<<endl;
    }
    return 0;
}
```

4.27.7 汉语翻译

1. 题目

问 题 的 根

给定正整数 B 和 N，找出一个整数 A，使得 A^N 最靠近 B（A 是逼近 B 的第 N 个根）。注意 A^N 可能小于、等于或者大于 B。

2. 输入描述

输入数据由一对或多对 B 和 N 的值组成。每对 B 和 N 打印在一行上，中间用一个空格隔开。B 和 N 都为 0 的那行表示输入的结束。B 的值是从 1 到 1 000 000 的闭区间，N 的值是 1 到 9 的闭区间。

3. 输出描述

对于输入数据中的每对 B 和 N，输出上文中定义的 A 在一行上。

4. 输入/输出样例

输入样例	输出样例
4 3	1
5 3	2
27 3	3
750 5	4
1000 5	4
2000 5	4
3000 5	5
1000000 5	16
0 0	

4.28 Magic Square

4.28.1 链接地址

http://www.realoj.com/网上第 112 题

4.28.2 时空限制

Time Limit: 1000 ms   Resident Memory Limit: 1024 KB   Output Limit: 1024 B

4.28.3 题目内容

In recreational mathematics, a magic square of *n*-degree is an arrangement of n^2 numbers, distinct integers, in a square, such that the *n* numbers in all rows, all columns, and both diagonals sum to the same constant. For example, the picture below shows a 3-degree magic square using the integers of 1 to 9.

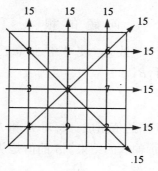

3-degree magic square

Given a finished number square, we need you to judge whether it is a magic square.

Input

The input contains multiple test cases.

The first line of each case stands an only integer N ($0 < N < 10$), indicating the degree of the number square and then N lines follows, with N positive integers in each line to describe the number square. All the numbers in the input do not exceed 1000.

A case with $N = 0$ denotes the end of input, which should not be processed.

Output

For each test case, print "Yes" if it's a magic square in a single line, otherwise print "No".

Sample Input

```
2
1 2
3 4
2
4 4
4 4
3
8 1 6
3 5 7
4 9 2
4
```

```
16 9 6 3
5 4 15 10
11 14 1 8
2 7 12 13
0
```

Sample Output

```
No
No
Yes
Yes
```

4.28.4 题目来源

Zhejiang University Local Contest 2007 (Author: JIANG, Hao)

4.28.5 解题思路

本题要求判断一个矩阵的每行、每列和两条对角线的和是否相等，如果相等，那么它就是一个幻方。但这还有一个前提，即这个矩阵中的所有数字不会重复。

这里特别指出一点，题目中举的 3 度幻方的例子中，每个数字正好是在 1～9 之间，但这并不等于规定了矩阵元素必须位于 $1 \sim n^2$ 的闭区间中。实际上，输入描述中的这句话 "All the numbers in the input do not exceed 1000."（输入数据中所有的数不会超过 1 000），指出了幻方中的数字是在 1～1 000 之间的闭区间中。

从本题我们可以得到一个启示，题目中给出的例子只是全部测试案例中的特例而已，而不是全貌。要知道全貌，就要好好看懂各个边界条件。

4.28.6 参考答案

```
#include <fstream>
#include <iostream>
#include <set>
using namespace std;
int main(int argc, char * argv[])
{
    ifstream cin("aaa.txt");
    int matrix[9][9];
    int line[9],column[9];//每行和每列的和
    int a,b;//两条对角线的和
    set<int>s;
    int n,i,j;
    while(cin>>n)
    {
        if(n==0)break;
        //读入矩阵
        for(i=0;i<n;i++)
        {
```

```
            for(j=0;j<n;j++)
            {
                cin>>matrix[i][j];
            }
        }
        s.clear();//清空集合
        for(i=0;i<n;i++)
        {
            for(j=0;j<n;j++)
            {
                //把每个元素放到集合中,重复数据不会插入
                //All the numbers in the input do not exceed 1000
                //说明矩阵块中的元素的范围是1~1000,而不限于在1~n² 范围内
                s.insert(matrix[i][j]);
            }
        }
        if(s.size()!=n*n)
        {
            cout<<"No"<<endl;
            continue;
        }
        //计算每行、每列、两条对角线的和
        a=0;
        b=0;
        for(i=0;i<n;i++)
        {
            line[i]=0;
            column[i]=0;
        }
        for(i=0;i<n;i++)
        {
            for(j=0;j<n;j++)
            {
                line[i]+=matrix[i][j];
                column[j]+=matrix[i][j];
                if(i==j)a=a+matrix[i][j];
                if(i+j==n-1)b=b+matrix[i][j];
            }
        }
        //判断每行、每列、两条对角线的和是否完全相等
        s.clear();//清空集合
        s.insert(a);
        s.insert(b);
        for(i=0;i<n;i++)
        {
            s.insert(line[i]);
            s.insert(column[i]);
        }
        if(s.size()!=1)cout<<"No"<<endl;
```

```
        else cout<<"Yes"<<endl;
    }
    return 0 ;
}
```

4.28.7　汉语翻译

1. 题目

幻　　方

在娱乐性的数学里，一个 n 度幻方共有 n^2 个数字的排列，它们都是不同的整数，在一个方块中，n 个数字在所有行、所有列和所有对角线中的和都相同。例如，下图显示了一个使用了整数 1~9 的 3 度幻方。

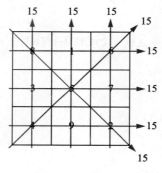

3 度幻方

给定一个填充好数字的方形，我们需要你来判断它是否是一个幻方。

2. 输入描述

输入数据包含多个测试案例。

每个测试案例的第一行是一个整数 N（$0 < N < 10$），表示该数字方形的度数，然后下面是 N 行，每行有 N 个正整数描述了这个数字方形。输入数据中的所有数字不会超过 1 000。$N=0$ 的测试案例表示输入的结束，不要处理它。

3. 输出描述

对于每个测试案例，如果它是幻方，打印一个"Yes"在一行上，否则，打印"No"。

4. 输入样例

```
2
1 2
3 4
2
4 4
4 4
3
8 1 6
```

```
3 5 7
4 9 2
4
16 9 6 3
5 4 15 10
11 14 1 8
2 7 12 13
0
```

5. 输出样例

```
No
No
Yes
Yes
```

4.29 Semi-Prime

4.29.1 链接地址

http://www.realoj.com/ 网上第 113 题

4.29.2 时空限制

Time Limit: 1000 ms   Resident Memory Limit: 1024 KB   Output Limit: 1024 B

4.29.3 题目内容

Prime Number Definition

An integer greater than one is called a prime number if its only positive divisors (factors) are one and itself. For instance, 2, 11, 67, 89 are prime numbers but 8, 20, 27 are not.

Semi-Prime Number Definition

An integer greater than one is called a semi-prime number if it can be decompounded to TWO prime numbers. For example, 6 is a semi-prime number but 12 is not.

Your task is just to determinate whether a given number is a semi-prime number.

Input

There are several test cases in the input. Each case contains a single integer N ($2 \leq N \leq 1\,000\,000$).

Output

One line with a single integer for each case. If the number is a semi-prime number, then output "Yes", otherwise "No".

Sample Input

```
3
4
```

```
6
12
```

Sample Output

```
No
Yes
Yes
No
```

4.29.4 题目来源

Zhejiang University Local Contest 2006, Preliminary (Author: LIU, Yaoting)

4.29.5 解题思路

本题判断一个数是否是半素数，由于数据量大，最易出现超时的错误了。

素数：只能被本身和 1 整除的数，1 不是，如 2，3，5，7，11。

半素数：只有三个因子——1 和两个素数因子，这两个素数可以是相同的，如 $4=1\times2\times2$。

本题解法：

（1）建立[2, 500 000]范围内的素数表，放在向量里；

（2）建立[2, 1 000 000]范围内的半素数表，放在集合里，集合 set 是平衡检索二叉树，检索速度最快；

（3）读入数据，直接在平衡检索二叉树里检索。

4.29.6 参考答案

```
#include <iostream>
#include <vector>
#include <set>
#include <cmath>
using namespace std;
//建立全局向量，用来保存素数
vector<int> v;
//在全局内存中定义全局集合容器，用来保存半素数
//集合是平衡检索二叉树，搜索速度最快
set<int> s;
//建立[a,b]范围内的素数表
void pt(int a,int b)
{
    for(int i=a;i<=b;i++)
    {
        //2 是素数，这里清除 2 的倍数
        if(i!=2 && i%2==0)continue;
        //清除素数的倍数
        for(int j=3;j*j<=i;j+=2)
        {
```

```
            if(i%j==0)goto RL;
        }
        //放入向量，不排序
        v.push_back(i);
RL:continue;
    }
}
int main(int argc, char * argv[])
{
    ifstream cin("aaa.txt");
    //建立[2,500000]范围内的素数表
    pt(2,500000);
    //建立 2~1000000 范围内的半素数表，两个素数相乘
    int i,j,p;
    for(i=0;i<v.size();i++)
    {
        for(j=0;j<v.size();j++)
        {
            p=v[i]*v[j];
            if(p<1000000)
                s.insert(p);
            else
                break;
        }
    }
    //读入数据，在半素数表中查找，看是否存在该表
    int n;
    set<int>::iterator it;
    while(cin>>n)
    {
        it=s.find(n);
        if(it!=s.end())
            cout<<"Yes"<<endl;
        else
            cout<<"No"<<endl;
    }
    return 0;
}
```

4.29.7 汉语翻译

1. 题目

半 素 数

素数的定义

素数是指大于1，且只能被本身和1整除的正整数（只有1和本身两个正因子）。例如，2，11，67，89都是素数，但8，20，27都不是素数。

半素数的定义

如果一个数能分解为两个素数的乘积，且大于 1，那么，这个数就是半素数。例如，6=2×3，2 和 3 都是素数，那么，6 就是半素数。12 就不是半素数。

你的任务就是判断一个给定的数字是否是一个半素数。

2. 输入描述

输入数据中有多个测试案例。每个案例只包含一个整数 N（2≤N≤1 000 000）。

3. 输出描述

一个测试案例输出一行。如果这个数是半素数，则直接输出"Yes"，否则，输出"No"。

4. 输入样例

```
3
4
6
12
```

5. 输出样例

```
No
Yes
Yes
No
```

4.30 Beautiful Number

4.30.1 链接地址

http://www.realoj.com/网上第 114 题

4.30.2 时空限制

Time Limit: 1000 ms　　Resident Memory Limit: 1024 KB　　Output Limit: 1024 B

4.30.3 题目内容

Mike is very lucky, as he has two beautiful numbers, 3 and 5. But he is so greedy that he wants infinite beautiful numbers. So he declares that any positive number which is dividable by 3 or 5 is beautiful number.

Given you an integer N ($1 \leq N \leq 100\,000$), could you please tell mike the Nth beautiful number?

Input

The input consists of one or more test cases. For each test case, there is a single line containing an integer N.

Output

For each test case in the input, output the result on a line by itself.

Sample Input

```
1
2
3
4
```

Sample Output

```
3
5
6
9
```

4.30.4 题目来源

Zhejiang University Local Contest 2007, Preliminary (Author: MAO, Yiqiang)

4.30.5 解题思路

本题很有趣，也很容易被迷惑，题目要求输出第 N（$1 \leqslant N \leqslant 100\,000$）个漂亮的数字，很显然，本题要求的最大的漂亮的数字不是 100 000，而是第 100 000 个漂亮的数字。

先建立第 1 到第 100 000 个漂亮的数字表，然后，直接查表就行。

采用映照容器是比较快的，它是平衡检索二叉树，所以，检索速度很快。

4.30.6 参考答案

```cpp
#include <fstream>
#include <iostream>
#include <map>
using namespace std;
//建立全局映照容器保存漂亮数
//映照容器是黑白树，检索快
map<int,int> m;
int main(int argc, char * argv[])
{
    ifstream cin("aaa.txt");
    //建立第 1 到第 100000 个漂亮数表
    int i=0;
    int p=0;
    while(1)
```

```
    {
        i++;
        if(i%3==0 || i%5==0)
        {
            p++;
            if(p>100000)break;//到了第100000个漂亮数就退出
            m[p]=i;
        }
    }
    //读入数据,并从映照容器中检索
    int n;
    map<int,int>::iterator it;
    while(cin>>n)
    {
        cout<<m[n]<<endl;
    }
    return 0;
}
```

4.30.7 汉语翻译

1. 题目

漂亮的数字

麦克非常幸运,因为他有两个幸运数字 3 和 5。但他非常贪婪,他想列出所有的漂亮数字。所以,他声明所有能被 3 或 5 整除的正整数都算是漂亮数字。

给你任何一个整数 N($1 \leqslant N \leqslant 100\,000$),你能告诉麦克第 N 个漂亮数字吗?

2. 输入

输入包含一个或多个测试案例。每个测试案例是一个整数 N,占一行。

3. 输出

对于输入数据中的每个测试案例,在单独一行上输出结果。

4. 输入样例

```
1
2
3
4
```

5. 输出样例

```
3
5
6
9
```

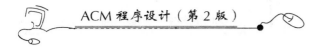

4.31 Phone List

4.31.1 链接地址

http://www.realoj.com/网上第 115 题

4.31.2 时空限制

Time Limit: 1000 ms   Resident Memory Limit: 1024 KB   Output Limit: 1024 B

4.31.3 题目内容

Given a list of phone numbers, determine if it is consistent in the sense that no number is the prefix of another. Let's say the phone catalogue listed these numbers:

- Emergency 911
- Alice 97 625 999
- Bob 91 12 54 26

In this case, it's not possible to call Bob, because the central would direct your call to the emergency line as soon as you had dialled the first three digits of Bob's phone number. So this list would not be consistent.

Input

The first line of input gives a single integer, $1 \leqslant t \leqslant 40$, the number of test cases. Each test case starts with n, the number of phone numbers, on a separate line, $1 \leqslant n \leqslant 10\,000$. Then follows n lines with one unique phone number on each line. A phone number is a sequence of at most ten digits.

Output

For each test case, output "YES" if the list is consistent, or "NO" otherwise.

Sample Input

```
2
3
911
97625999
91125426
5
113
12340
123440
12345
98346
```

Sample Output

NO
YES

4.31.4 题目来源

The 2007 Nordic Collegiate Programming Contest

4.31.5 解题思路

本题是查找一个串是否是另一个串的前缀，看似简单，如果直接去查找，那么，会出现超时错误。

其实，只要将读入的字符串由小到大排列，然后到下一个中去查找上一个即可，如果下一个串的最前面的子串是上一字符串，那么就终止。

参考答案中给出了C++输入输出 cin 和 cout 与 C 输入输出 scanf 和 printf 耗时量的比较，说明使用 C 输入输出十分快速。当然，在 ACM 竞赛中，如果不超时，谁先做对，谁就赢了。如果超时了，那么，得考虑是否是时间开销在频繁的输入输出上。其次，再考虑是否是算法过于低效。

4.31.6 参考答案

（1）用 cin>>s，耗时 2.09s。

```
#include <fstream>
#include <iostream>
#include <string>
#include <vector>
#include <algorithm>
using namespace std;
int main(int argc, char * argv[])
{
    ifstream cin("aaa.txt");
    vector<string> v;
    //char ch[10];
    string s;
    int n;
    cin>>n;//不理会总块数
    int i,j;
    while(cin>>n)
    {
        v.clear();
        for(i=0;i<n;i++)
        {
            cin>>s;
            //scanf("%s",&ch);
            //s=ch;
            v.push_back(s);
```

```
        }
        sort(v.begin(),v.end());
        for(i=0;i<n-1;i++)
        {
            if(v[i+1].find(v[i])==0)
            {
                cout<<"NO"<<endl;
                //printf("NO\n");
                goto RL;
            }
        }
        cout<<"YES"<<endl;
        //printf("YES\n");
RL:
        continue;
    }
    return 0;
}
```

(2) 采用 scanf 输入方式，耗时 0.63s。

```
#include <fstream>
#include <iostream>
#include <string>
#include <vector>
#include <algorithm>
using namespace std;
int main(int argc, char * argv[])
{
    ifstream cin("aaa.txt");
    vector<string> v;
    char ch[10];
    string s;
    int n;
    cin>>n;//不理会总块数
    int i,j;
    while(cin>>n)
    {
        v.clear();
        for(i=0;i<n;i++)
        {
            //cin>>s;
            scanf("%s",&ch);
            s=ch;
            v.push_back(s);
        }
        sort(v.begin(),v.end());
        for(i=0;i<n-1;i++)
        {
            if(v[i+1].find(v[i])==0)
            {
```

```
                    cout<<"NO"<<endl;
                    //printf("NO\n");
                    goto RL;
                }
            }
            cout<<"YES"<<endl;
            //printf("YES\n");
RL:
            continue;
    }
    return 0;
}
```

（3）采用 scanf 输入，再采用 printf 输出，耗时 0.63s。

```
#include <fstream>
#include <iostream>
#include <string>
#include <vector>
#include <algorithm>
using namespace std;
int main(int argc, char * argv[])
{
    ifstream cin("aaa.txt");
    vector<string> v;
    char ch[10];
    string s;
    int n;
    cin>>n;//不理会总块数
    int i,j;
    while(cin>>n)
    {
        v.clear();
        for(i=0;i<n;i++)
        {
            //cin>>s;
            scanf("%s",&ch);
            s=ch;
            v.push_back(s);
        }
        sort(v.begin(),v.end());
        for(i=0;i<n-1;i++)
        {
            if(v[i+1].find(v[i])==0)
            {
                //cout<<"NO"<<endl;
                printf("NO\n");
                goto RL;
            }
        }
```

```
            //cout<<"YES"<<endl;
            printf("YES\n");
RL:
            continue;
        }
        return 0;
}
```

4.31.7 汉语翻译

1. 题目

电 话 列 表

给定一列电话号码，根据没有号码是其他号码的前缀，看看它是否符合要求。让我们来看看下面这列电话号码：

-- 紧急呼叫 911

-- 艾利丝 97 625 999

-- 鲍博 91 12 54 26

在这个案例里，就不可能打电话给鲍博，因为当你拨了鲍博电话前三位数字的时候，电话服务中心将拨通紧急电话。所以，这个电话列表不符合要求。

2. 输入描述

输入数据的第一行给出了一个整数，$1 \leqslant t \leqslant 40$，表示整个测试案例的个数。每个测试案例的开头一行是 n，表示电话号码的个数，$1 \leqslant n \leqslant 10\,000$。接下去的 n 行中，每行都是一个不相同的电话号码。每个电话号码都是一个至多 10 个数字的系列。

3. 输出描述

对于每个测试案例，如果它是符合要求的，那么输入"YES"，否则，输出"NO"。

4. 输入样例

```
2
3
911
97625999
91125426
5
113
12340
123440
12345
98346
```

5. 输出样例

```
NO
YES
```

4.32 Calendar

4.32.1 链接地址

http://www.realoj.com/网上第 116 题

4.32.2 时空限制

Time Limit: 1000 ms   Resident Memory Limit: 1024 KB   Output Limit: 1024 B

4.32.3 题目内容

A calendar is a system for measuring time, from hours and minutes, to months and days, and finally to years and centuries. The terms of hour, day, month, year and century are all units of time measurements of a calender system.

According to the Gregorian calendar, which is the civil calendar in use today, years evenly divisible by 4 are leap years, with the exception of centurial years that are not evenly divisible by 400. Therefore, the years 1700, 1800, 1900 and 2100 are not leap years, but 1600, 2000, and 2400 are leap years.

Given the number of days that have elapsed since January 1, 2000 A.D, your mission is to find the date and the day of the week.

Input

The input consists of lines each containing a positive integer, which is the number of days that have elapsed since January 1, 2000 A.D. The last line contains an integer -1, which should not be processed. You may assume that the resulting date won't be after the year 9999.

Output

For each test case, output one line containing the date and the day of the week in the format of "YYYY-MM-DD DayOfWeek", where "DayOfWeek" must be one of "Sunday", "Monday", "Tuesday", "Wednesday", "Thursday", "Friday" and "Saturday".

Sample Input

```
1730
1740
1750
1751
-1
```

Sample Output

```
2004-09-26 Sunday
2004-10-06 Wednesday
```

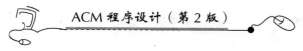

```
2004-10-16 Saturday
2004-10-17 Sunday
```

4.32.4 题目来源

Asia 2004, Shanghai (Mainland China), Preliminary

4.32.5 解题思路

本题给出自公元 2000 年 1 月 1 日消逝的天数，你的任务是要找出这天的日期和星期。这种频繁计算的程序，很容易出现超时错误，而且程序编制上太困难。

最好的办法是先用向量建立一个表，把自 2000 年 1 月 1 日到 9999 年 12 月 31 日之间的日期与星期信息放在表中，最后，通过查表的方式把日期信息打印出来。虽然建表的时间会开销多一点，但表建好了，再查就十分快速了。

这里牵涉到日期的一些常识。公历中，一年有 12 个月；1、3、5、7、8、10、12 月为大月，每月有 31 天；4、6、9、11 月为小月，每月有 30 天；如果是闰年，2 月有 29 天，否则，2 月只有 28 天。

判断一个年份是否是闰年，只要符合以下两个条件之一就是闰年：该年份能被 4 整数，但不能被 100 整除；该年份能被 400 整除。

4.32.6 参考答案

```cpp
#include <fstream>
#include <iostream>
#include <vector>
using namespace std;
//以下为阳历年月日的常识
//一年有 12 个月
//1 3 5 7 8 10 12 月为大月，每月有 31 天
//4 6 9 11 月为小月，每月有 30 天
//如果是闰年，2 月有 29 天，否则，2 月只有 28 天
//定义打印日期结构
struct Info{
    short int year;
    short int month;
    short int day;
    short int week;
};
//定义向量来存储日期信息表
vector<Info> v;
bool RR(int year)
{
    //符合下面两个条件之一，就是闰年
    //（1）能被 4 整除，但不能被 100 整除
    //（2）能被 4 整除，又能被 400 整除
    if((year%4==0 && year%100!=0) || (year%400==0))
```

```
            return true;
    else
            return false;
}
int main(int argc, char * argv[])
{
    ifstream cin("aaa.txt");
    int i,j,k;
    int p=0;
    Info info;
    int flag=0;//闰年标志
    int week=5;
    //建立2000-01-01到9999-12-31的日期计日表
    for(i=2000;i<=9999;i++)//年份
    {
        flag=RR(i);
        for(j=1;j<=12;j++)//月份
        {
            for(k=1;k<=31;k++)//日
            {
                //1 3 5 7 8 10 12月为大月,每月有31天
                if((j==1) || (j==3) || (j==5) || (j==7) || (j==8) || (j==10)
                    || (j==12))
                    p++;
                //4 6 9 11月为小月,每月有30天
                else if((j==4) || (j==6) || (j==9) || (j==11))
                {
                    if(k!=31)p++;
                    else
                        break;//跳到下一个月
                }
                //如果是闰年,2月有29天,否则,2月只有28天
                else if(j==2)
                {
                    if(flag)//是闰年
                    {
                        if(k!=30 && k!=31)p++;
                        else
                            break;//跳到下一个月
                    }
                    else//不是闰年
                    {
                        if(k!=29 && k!=30 && k!=31)p++;
                        else
                            break;//跳到下一个月
                    }
                }
                info.year=i;
                info.month=j;
```

```
                    info.day=k;
                    week++;
                    if(week==8)week=1;
                    info.week=week;
                    v.push_back(info);
                }
            }
        }
        int n;
        while(cin>>n)
        {
            if(n==-1)break;
            else
            {
                cout<<v[n].year<<"-";
                if(v[n].month<10)cout<<"0"<<v[n].month<<"-";
                else
                    cout<<v[n].month<<"-";
                if(v[n].day<10)cout<<"0"<<v[n].day<<" ";
                else
                    cout<<v[n].day<<" ";
                if(v[n].week==1)
                    cout<<"Monday"<<endl;
                else if(v[n].week==2)
                    cout<<"Tuesday"<<endl;
                else if(v[n].week==3)
                    cout<<"Wednesday"<<endl;
                else if(v[n].week==4)
                    cout<<"Thursday"<<endl;
                else if(v[n].week==5)
                    cout<<"Friday"<<endl;
                else if(v[n].week==6)
                    cout<<"Saturday"<<endl;
                else if(v[n].week==7)
                    cout<<"Sunday"<<endl;
            }
        }
        return 0;
    }
```

4.32.7 汉语翻译

1. 题目

日　历

日历是一种测量时间的系统，从小时和分钟，到月和日，最后到年和世纪。小时、天、月、年和世纪这些术语都是日历系统中的时间测量单位。

根据公历，也就是今天的民用历，能被4整除的年份是闰年，当然，不能被400整除的世纪年要除外。因此，1700，1800，1900 和 2100 这几个年份不是闰年，但 1600，2000 和 2400 这几个年份则是闰年。

给出自公元 2000 年 1 月 1 日起消逝的天数，你的任务是要找出这天的日期和星期。

2. 输入描述

输入的每一行是一个正整数，表示自公元 2000 年 1 月 1 日起消逝的天数。最后一行是一个整数 -1 表示输入的结束，你不要去处理。所有的年份不会超过 9999。

3. 输出描述

对于每个测试案例，输入一个日期和这天的星期，格式是"YYYY-MM-DD DayOfWeek"，这里"DayOfWeek"必须是"Sunday"（星期天）、"Monday"（星期一）、"Tuesday"（星期二）、"Wednesday"（星期三）、"Thursday"（星期四）、"Friday"（星期五）和"Saturday"（星期六）中的一个。

4. 输入样例

```
1730
1740
1750
1751
-1
```

5. 输出样例

```
2004-09-26 Sunday
2004-10-06 Wednesday
2004-10-16 Saturday
2004-10-17 Sunday
```

4.33 · No Brainer

4.33.1 链接地址

http://www.realoj.com/ 网上第 117 题

4.33.2 时空限制

Time Limit: 1000 ms   Resident Memory Limit: 1024 KB   Output Limit: 1024 B

4.33.3 题目内容

Zombies love to eat brains. Yum.

Input

The first line contains a single integer *n* indicating the number of data sets.

The following *n* lines each represent a data set. Each data set will be formatted according to the following description:

A single data set consists of a line "*X Y*", where *X* is the number of brains the zombie eats and *Y* is the number of brains the zombie requires to stay alive.

Output

For each data set, there will be exactly one line of output. This line will be "MMM BRAINS" if the number of brains the zombie eats is greater than or equal to the number of brains the zombie requires to stay alive. Otherwise, the line will be "NO BRAINS".

Sample Input

```
3
4 5
3 3
4 3
```

Sample Output

```
NO BRAINS
MMM BRAINS
MMM BRAINS
```

4.33.4 题目来源

South Central USA 2004

4.33.5 解题思路

本题虽然简单，但很受启发。简单在于它的程序实现，很受启发则在于它要调动我们的脑子去思考，去解决问题。

僵尸吃人，要吃固定数量的人才能存活下来。如果吃的人比需要的数量少，当然，它就活不下来了，否则，就能存活下来。所以，本题就是直接比较两个数，如果吃了的人的数量比需要吃的人数量少，则直接打印"**NO BRAINS**"，否则，打印"**MMM BRAINS**"即可。

4.33.6 参考答案

```cpp
#include <iostream>
#include <fstream>
using namespace std;
int main(int argc, char * argv[])
{
    ifstream cin("aaa.txt");
    int a,b;
    cin>>a;
    while(cin>>a>>b)
    {
        if(a<b)cout<<"NO BRAINS"<<endl;
```

```
        else
            cout<<"MMM BRAINS"<<endl;
    }
    return 0;
}
```

4.33.7 汉语翻译

1. 题目

不 费 脑 筋

僵尸爱吃人。味道好极了。

2. 输入描述

第一行是一个整数 n，表示测试案例的个数。

接下来的 n 行，每行表示一个数据集。每个数据集的格式如下：

每个数据集是两个数"$X\ Y$"，它们在一行上，这里，X 是僵尸吃的人的个数，Y 是僵尸活下来需要吃的人个数。

3. 输出描述

对于每个数据集，输出一行信息。如果僵尸吃的人数目大于或等于他活下来需要吃的人数目，打印一行"MMM BRAINS"；否则，打印一行"NO BRAINS"。

4. 输入样例

```
3
4 5
3 3
4 3
```

5. 输出样例

```
NO BRAINS
MMM BRAINS
MMM BRAINS
```

4.34 Quick Change

4.34.1 链接地址

http://www.realoj.com/网上第 118 题

4.34.2 时空限制

Time Limit: 1000 ms   Resident Memory Limit: 1024 KB   Output Limit: 1024 B

4.34.3 题目内容

J. P. Flathead's Grocery Store hires cheap labor to man the checkout stations. The people he hires (usually high school kids) often make mistakes making change for the customers. Flathead, who's a bit of a tightwad, figures he loses more money from these mistakes than he makes; that is, the employees tend to give more change to the customers than they should get.

Flathead wants you to write a program that calculates the number of quarters ($0.25), dimes ($0.10), nickels ($0.05) and pennies ($0.01) that the customer should get back. Flathead always wants to give the customer's change in coins if the amount due back is $5.00 or under. He also wants to give the customers back the smallest total number of coins. For example, if the change due back is $1.24, the customer should receive 4 quarters, 2 dimes, 0 nickels, and 4 pennies.

Input

The first line of input contains an integer N which is the number of datasets that follow. Each dataset consists of a single line containing a single integer which is the change due in cents, C ($1 \leq C \leq 500$).

Output

For each dataset, print out the dataset number, a space, and the string:

Q QUARTER(S), D DIME(S), n NICKEL(S), P PENNY(S)

Where Q is he number of quarters, D is the number of dimes, n is the number of nickels and P is the number of pennies.

Sample Input

```
3
124
25
194
```

Sample Output

```
1 4 QUARTER(S), 2 DIME(S), 0 NICKEL(S), 4 PENNY(S)
2 1 QUARTER(S), 0 DIME(S), 0 NICKEL(S), 0 PENNY(S)
3 7 QUARTER(S), 1 DIME(S), 1 NICKEL(S), 4 PENNY(S)
```

4.34.4 题目来源

Greater New York Regional 2006

4.34.5 解题思路

本题是把美分数转化为 QUARTER，DIME，NICKEL，PENNY，实质上是整除和取余计算。

本题要注意输出格式，输出中的字符串，应当从网页中复制，这样，确保输出格式不会出错。

4.34.6 参考答案

```
#include <fstream>
#include <iostream>
using namespace std;
int main(int argc, char * argv[])
{
    ifstream cin("aaa.txt");
    int q,d,n,p;
    int m;
    cin>>m;//不理会第一项数据
    int i=0;
    while(cin>>m)
    {
            q=m/25;
            m=m%25;

            d=m/10;
            m=m%10;

            n=m/5;
            m=m%5;

            p=m;
            i++;
            cout<<i<<" "<<q<<" QUARTER(S), "<<d<<" DIME(S), "<<n<<" NICKEL(S), "<<p<<" PENNY(S)"<<endl;
    }
    return 0;
}
```

4.34.7 汉语翻译

1. 题目

快 速 找 零

　　J. P. Flathead 的杂货店在收银台上雇佣廉价的劳动力。他雇的人（常常是高中的孩子）常常找错钱给顾客。Flathead 是个吝啬鬼，估计他找错的钱比他挣的还要多，也就是说，雇员往往找更多的钱给客户。

　　Flathead 想你编写一个程序来计算他应该找多少 quarters（$0.25）、dimes（$0.10）、nickels（$0.05）和 pennies（$0.01）给顾客。如果要找的钱是 5.00 美元或者更少，Flathead 尽量找给他们硬币。他也想找给顾客最少数量的硬币。比如，要找给顾客的钱是 1.24 美元，那么，顾客应当拿到 4 quarters、2 dimes、0 nickels 和 4 pennies。

2. 输入描述

输入数据的第一行包含一个整数 N，表示下面数据集的个数。每个数据集是由一个整数组成，占一行，这个整数代表 cents（美分），C（$1 \leqslant C \leqslant 500$）。

3. 输出描述

对于每个数据集，打印打出数据集的序号，一个空格和下面这串字符串：
Q QUARTER(S), D DIME(S), n NICKEL(S), P PENNY(S)
这里，Q 是 quarters 数目，D 是 dimes 数目，n 是 nickels 数目，而 P 是 pennies 数目。

4. 输入样例

```
3
124
25
194
```

5. 输出样例

```
1 4 QUARTER(S), 2 DIME(S), 0 NICKEL(S), 4 PENNY(S)
2 1 QUARTER(S), 0 DIME(S), 0 NICKEL(S), 0 PENNY(S)
3 7 QUARTER(S), 1 DIME(S), 1 NICKEL(S), 4 PENNY(S)
```

4.35 Total Amount

4.35.1 链接地址

http://www.realoj.com/网上第 119 题

4.35.2 时空限制

Time Limit: 1000 ms   Resident Memory Limit: 1024 KB   Output Limit: 1024 B

4.35.3 题目内容

Given a list of monetary amounts in a standard format, please calculate the total amount. We define the format as follows:

(1) The amount starts with "$".

(2) The amount could have a leading "0" if and only if it is less then 1.

(3) The amount ends with a decimal point and exactly 2 following digits.

(4) The digits to the left of the decimal point are separated into groups of three by commas (a group of one or two digits may appear on the left).

Input

The input consists of multiple tests. The first line of each test contains an integer N ($1 \leqslant N \leqslant 10\,000$) which indicates the number of amounts. The next N lines contain N amounts. All amounts and the total amount are between $0.00 and $20 000 000.00, inclusive. *N*=0 denotes the end of input.

Output

For each input test, output the total amount.

Sample Input

```
2
$1,234,567.89
$9,876,543.21
3
$0.01
$0.10
$1.00
0
```

Sample Output

```
$11,111,111.10
$1.11
```

4.35.4　题目来源

Zhejiang Provincial Programming Contest 2005 (Author: ZHANG, Zheng)

4.35.5　解题思路

本题本质上是加法题，看似简单，实际上处理起来需要很强的基本功，综合性强，包括字符串处理、进位处理、反转容器算法、输出格式处理。

在算法上没有什么窍门，按照处理顺序，一行一行编程就好了，本题比较综合，很能锻炼编程的基本功。

4.35.6　参考答案

```cpp
#include <fstream>
#include <iostream>
#include <string>
#include <algorithm>
#include <map>
using namespace std;
int main(int argc, char * argv[])
{
    ifstream cin("aaa.txt");
    string sa,sb,t;
    int pa,pb,pc;//pc=pa+pb;
    //定义映照容器
    map<char,int>m;
    m['0']=0;
```

```
m['1']=1;
m['2']=2;
m['3']=3;
m['4']=4;
m['5']=5;
m['6']=6;
m['7']=7;
m['8']=8;
m['9']=9;
//定义映照容器
map<int,char>mm;
mm[0]='0';
mm[1]='1';
mm[2]='2';
mm[3]='3';
mm[4]='4';
mm[5]='5';
mm[6]='6';
mm[7]='7';
mm[8]='8';
mm[9]='9';
int n,i,j;
//进位标记，0表示无进位，1表示有进位
int flag=0;
while(cin>>n)
{
    if(n==0)break;
    for(i=0;i<n;i++)
    {
        cin>>sb;
        //删除"$"符号
        sb.erase(0,1);
        //删除","符号
        t="";
        for(j=0;j<sb.size();j++)
        {
            if(sb[j]!=',')t+=sb[j];
        }
        sb=t;
        //反转字符
        reverse(sb.begin(),sb.end());
        //删除"."符号
        sb.erase(2,1);
        //读入的是第一个字符串
        if(i==0)sa=sb;
        //sa + sb
        else
```

```
            {
                //进位标志设为 0, 即无进位
                flag=0;
                //把长的字符串放到 sa 里
                if(sa.size()<sb.size())
                {
                    t=sa;sa=sb;sb=t;
                }
                for(j=0;j<sa.size();j++)
                {
                    pa=m[sa[j]];
                    if(j>=sb.size())pb=0;
                    else pb=m[sb[j]];
                    pc=pa + pb + flag;
                    if(pc>9)
                    {
                        pc=pc-10;//本位
                        flag=1;//有进位
                    }
                    else flag=0;//进位标记清 0
                    sa[j]=mm[pc];
                }
                //如果有进位, 则需再加一位
                if(flag==1)sa+="1";
            }
        }
        //加入小数点和逗号
        t="";
        for(i=0;i<sa.size();i++)
        {
            t=t+sa[i];
            //加"."号
            if(i==1)t=t+".";
            //加","号
            if(i!=1 && (i-1)%3==0 && i!=(sa.size()-1))
                t=t+",";
        }
        sa=t;
        //反向输出结果
        cout<<"$";
        for(i=sa.size()-1;i>=0;i--)
            cout<<sa[i];
        cout<<endl;
    }
    return 0;
}
```

4.35.7 汉语翻译

1. 题目

总　和

给出一列标准格式的货币数字，请计算出它们的总和。

我们定义货币使用下面的格式：

（1）货币以"$"符号打头。
（2）如果货币小于1，则整数部分是一个"0"。
（3）货币都有两位小数。
（4）货币的小数点左边的每三位都用逗号隔开为一组（最左边的一组可能是一位或两位数字）。

2. 输入描述

输入数据包含多个测试案例。每个测试案例的第一行是一个整数 N（$1 \leqslant N \leqslant 10\,000$），表示该测试案例中的数字的个数。接下来的 N 行包含 N 个数字。每个数字和所有数字的和在 $0.00 到 $20\,000\,000.00 的闭区间中。$N=0$ 表示输入的结束。

3. 输出描述

对于每个测试案例，输出货币的总和。

4. 输入样例

```
2
$1,234,567.89
$9,876,543.21
3
$0.01
$0.10
$1.00
0
```

5. 输出样例

```
$11,111,111.10
$1.11
```

4.36　Electrical Outlets

4.36.1 链接地址

http://www.realoj.com/网上第120题

4.36.2 时空限制

Time Limit: 1000 ms   Resident Memory Limit: 1024 KB   Output Limit: 1024 B

4.36.3 题目内容

Roy has just moved into a new apartment. Well, actually the apartment itself is not very new, even dating back to the days before people had electricity in their houses. Because of this, Roy's apartment has only one single wall outlet, so Roy can only power one of his electrical appliances at a time.

Roy likes to watch TV as he works on his computer, and to listen to his HiFi system (on high volume) while he vacuums, so using just the single outlet is not an option. Actually, he wants to have all his appliances connected to a powered outlet, all the time. The answer, of course, is power strips, and Roy has some old ones that he used in his old apartment. However, that apartment had many more wall outlets, so he is not sure whether his power strips will provide him with enough outlets now.

Your task is to help Roy compute how many appliances he can provide with electricity, given a set of power strips. Note that without any power strips, Roy can power one single appliance through the wall outlet. Also, remember that a power strip has to be powered itself to be of any use.

Input

Input vill start with a single integer $1 \leq N \leq 20$, indicating the number of test cases to follow. Then follow N lines, each describing a test case. Each test case starts with an integer $1 \leq K \leq 10$, indicating the number of power strips in the test case. Then follow, on the same line, K integers separated by single spaces, $O_1 O_2 \cdots O_k$, where $2 \leq O_i \leq 10$, indicating the number of outlets in each power strip.

Output

Output one line per test case, with the maximum number of appliances that can be powered.

Sample Input

```
3
3 2 3 4
10 4 4 4 4 4 4 4 4 4 4
4 10 10 10 10
```

Sample Output

```
7
31
37
```

4.36.4　题目来源

Nordic 2005

4.36.5　解题思路

本题把题意看懂了，就比较简单，否则，猜题意是比较困难的。

题目说，墙上只有一个电源插口，但手头有很多插件板，问最多同时能接多少电器？

看懂了题意，问题就变得简单了，一个插件板需要使用一个插口，否则，该插件板本身也没有电。

4.36.6　参考答案

```
#include <fstream>
#include <iostream>
#include <vector>
using namespace std;
int main(int argc, char * argv[])
{
    ifstream cin("aaa.txt");
    int n;
    vector<int> v;
    int i,num,sum;
    cin>>n;//不理会第一个数字
    while(cin>>n)
    {
        v.clear();//清空向量
        sum=0;//清空
        for(i=0;i<n;i++)
        {
            cin>>num;
            v.push_back(num);
        }
        //计算可用来插入其他电器的插口数量
        for(i=0;i<v.size();i++)
        {
            if(i!=v.size()-1)//不是最后一个插件板
                sum=sum+v[i]-1;
            else
                sum=sum+v[i];
        }
        //输出提供插口数量
        cout<<sum<<endl;
    }
    return 0;
}
```

4.36.7 汉语翻译

1. 题目

电源插座

Roy 刚搬到一间新公寓，实际上公寓本身并不很新，甚至可以追溯到人们在这座公寓里用电之前的那些日子。正因为这样，Roy 的房间里只有一个墙上电源插座。所以，Roy 在某一时刻，只能让他的一个电器插电。

Roy 喜欢玩电脑的时候同时看电视，还要在用吸尘器打扫房间的时候听 HiFi 系统（高音），所以，只有一个插口肯定不行。实际上，他需要把他的所有电器都一直插上电源。当然，得使用电源插件板，Roy 有几块曾经在老房间里使用过的电源插件板。然而，原先那间房间里有很多电源插口，所以，他不能确定现在这几块电源插件板是否还够用。

你的任务是帮助 Roy 计算出他的插件板能提供多少个电器插入，插件板是给定的。记住，插件板自身也要有电源进入。

2. 输入描述

输入数据由一个整数打头，$1 \leqslant N \leqslant 20$，表示下面的测试案例的个数。接下来是 N 行，每行描述了一个测试案例。每个测试案例以一个整数 K 打头，$1 \leqslant K \leqslant 10$，表示这个测试案例中插件板的数量。然后，同一行中，跟着 K 个整数，中间用空格隔开，$O_1 O_2 \cdots O_k$，这里 $2 \leqslant O_i \leqslant 10$，表示每块插件板上插座的个数。

3. 输出描述

每个测试案例，对应地在一行上输出一个整数，这个整数表示该组插件板最多能插入的电器数量。

4. 输入样例

```
3
3 2 3 4
10 4 4 4 4 4 4 4 4 4 4
4 10 10 10 10
```

5. 输出样例

```
7
31
37
```

4.37 Speed Limit

4.37.1 链接地址

http://www.realoj.com/网上第 121 题

4.37.2 时空限制

Time Limit: 1000 ms   Resident Memory Limit: 1024 KB   Output Limit: 1024 B

4.37.3 题目内容

Bill and Ted are taking a road trip. But the odometer in their car is broken, so they don't know how many miles they have driven. Fortunately, Bill has a working stopwatch, so they can record their speed and the total time they have driven. Unfortunately, their record keeping strategy is a little odd, so they need help computing the total distance driven. You are to write a program to do this computation.

For example, if their log show

Speed in miles per hour	Total elapsed time in hours
20	2
30	6
10	7

This means they drove 2 hours at 20 miles per hour, then 6−2=4 hours at 30 miles per hour, then 7−6=1 hour at 10 miles per hour. The distance driven is then (2)(20) + (4)(30) + (1)(10) = 40 + 120 + 10 = 170 miles. Note that the total elapsed time is always since the beginning of the trip, not since the previous entry in their log.

Input

The input consists of one or more data sets. Each set starts with a line containing an integer n, $1 \leqslant n \leqslant 10$, followed by n pairs of values, one pair per line. The first value in a pair, s, is the speed in miles per hour and the second value, t, is the total elapsed time. Both s and t are integers, $1 \leqslant s \leqslant 90$ and $1 \leqslant t \leqslant 12$. The values for t are always in strictly increasing order. A value of −1 for n signals the end of the input.

Output

For each input set, print the distance driven, followed by a space, followed by the word "miles".

Example

Example Input	Example Output
3	170 miles
20 2	180 miles
30 6	90 miles
10 7	
2	
60 1	
30 5	
4	
15 1	
25 5	
30 3	
10 5	
-1	

4.37.4 题目来源

Mid-Central USA 2004

4.37.5 解题思路

本题是计算汽车所开的总里程,仔细读懂题目中举的例子,就能看懂怎么计算。

在竞赛中,很多同学都没有时间把题目精确翻译出来,这里,比较好的办法就是照着题目中举的例子,仔细分析一下,再把输入输出样例分析一下,然后在题目中找找数据输入输出的边界条件,一般也就把题意看得八九不离十了,这样节省了很多时间。尤其是有些题,题目很长,翻译起来很难,等你把题看懂了,其他队通过分析题例,早就把程序给编出来了。这也是竞赛中的一个取胜策略。

4.37.6 参考答案

```
#include <fstream>
#include <iostream>
using namespace std;
int main(int argc, char * argv[])
{
    ifstream cin("aaa.txt");
    int n,a,b,bb,sum,i;
    while(cin>>n)
    {
        if(n==-1)break;
        sum=0;
        bb=0;
        for(i=0;i<n;i++)
        {
            cin>>a>>b;
            if(i==0)
            {
```

```
                    bb=b;
                    sum=sum+a*bb;
                }
                else
                {
                    sum=sum+a*(b-bb);
                    bb=b;
                }
            }
            cout<<sum<<" miles"<<endl;
        }
        return 0;
    }
```

4.37.7 汉语翻译

1. 题目

限　　速

　　Bill 和 Ted 正在进行公路旅行。但在他们汽车里的里程表被打破了，所以他们不知道车开了多少英里。所幸的是，Bill 有一个可以工作的秒表，所以，他们可以记录他们开车的速度和总时间。不幸的是，他们保存记录的策略比较奇怪，所以他们需要帮助，计算出总的驾驶里程。你写一个程序来计算这项工作。

　　例如，如果是如下记录

行驶速度（英里/小时）	消耗的总时间（小时）
20	2
30	6
10	7

　　这就意味着以每小时 20 英里的速度开了 2 小时，然后，以每小时 30 英里的速度开了 6-2=4 小时，最后，以每小时 10 英里的速度开了 7-6=1 小时。那么，他们总共开了(2)(20) + (4)(30) + (1)(10) = 40 + 120 + 10 = 170 英里。注意，消耗的总时间总是从旅行的开始计算，而不是从前一条记录开始计算。

2. 输入描述

　　输入数据由多组测试案例组成。每个测试案例以一个整数 n 开始，$1 \leqslant n \leqslant 10$，跟着是 n 对值，一对值占一行。值对中，第一个值是 s，即开车速度；第二个值是 t，是总共消耗的时间。s 和 t 都是整数，$1 \leqslant s \leqslant 90$ 且 $1 \leqslant t \leqslant 12$。$t$ 值是一直增加的。当 n=-1 时，表示输入的结束。

3. 输出描述

　　对于每个输入数据集合，打印出开车总里程，再跟一个空格，再跟一个单词"miles"。

4. 输入/输出样例

输入样例	输出样例
3	170 miles
20 2	180 miles
30 6	90 miles
10 7	
2	
60 1	
30 5	
4	
15 1	
25 2	
30 3	
10 5	
-1	

4.38 Beat the Spread!

4.38.1 链接地址

http://www.realoj.com/网上第 122 题

4.38.2 时空限制

Time Limit: 1000 ms   Resident Memory Limit: 1024 KB   Output Limit: 1024 B

4.38.3 题目内容

Superbowl Sunday is nearly here. In order to pass the time waiting for the half-time commercials and wardrobe malfunctions, the local hackers have organized a betting pool on the game. Members place their bets on the sum of the two final scores, or on the absolute difference between the two scores.

Given the winning numbers for each type of bet, can you deduce the final scores?

Input

The first line of input contains n, the number of test cases. n lines follow, each representing a test case. Each test case gives s and d, non-negative integers representing the sum and (absolute) difference between the two final scores.

Output

For each test case, output a line giving the two final scores, largest first. If there are no such scores, output a line containing "impossible". Recall that football scores are always non-negative integers.

Sample Input

```
2
40 20
20 40
```

Sample Output

```
30 10
impossible
```

4.38.4 题目来源

University of Waterloo Local Contest 2005.02.05

4.38.5 解题思路

给出两个自然数的和与差，要求推算出这两个自然数（自然数是指 0 和正整数）。
解题分析如下。
（1）如果和小于差，则不可能（和必须要大于或等于差，分数必定是自然数，包括 0）。
（2）接下来这样推断两个分数 x 和 y：
设求解的数为 x 和 y，$x+y = a$；那么，$x-y = a$ 或 $x-y = -a$。这样就有以下两组方程。
第一组

$$\begin{cases} x+y=a \\ x-y=b \end{cases} \Rightarrow \begin{cases} x = \dfrac{a+b}{2} \\ y = \dfrac{a-b}{2} \end{cases}$$

第二组

$$\begin{cases} x+y=a \\ x-y=-b \end{cases} \Rightarrow \begin{cases} x = \dfrac{a-b}{2} \\ y = \dfrac{a+b}{2} \end{cases}$$

两组只要一组符合条件就行，这样，综合起来，要求的两个分数为 $\dfrac{a+b}{2}$ 和 $\dfrac{a-b}{2}$。当然，都要求绝对值的。
（3）求出的两个分数 x 和 y 都要是自然数（自然数包括 0）。

4.38.6 参考答案

```cpp
#include <fstream>
#include <iostream>
#include <cmath>
using namespace std;
int main(int argc, char * argv[])
```

```
{
    ifstream cin("aaa.txt");
    int n,x,y,a,b;
    cin>>n;//不理会第一个数字
    while(cin>>a>>b)
    {
        //和比差小，不成立
        if(a<b)
        {
            cout<<"impossible"<<endl;
            continue;
        }
        //分数只能是自然数（包括0）
        if((a+b)%2!=0 || abs(a-b)%2!=0)
        {
            cout<<"impossible"<<endl;
            continue;
        }
        x=abs((a+b)/2);
        y=abs((a-b)/2);
        if(x>y)
            cout<<x<<" "<<y<<endl;
        else
            cout<<y<<" "<<x<<endl;
    }
    return 0;
}
```

4.38.7 汉语翻译

1. 题目

Beat the Spread!

超级杯又来了，为了打发中场休息时间，大家就来下注最后的结果会如何。大家下注的目标为两队最后的分数或两队最后分数差的绝对值。

给你两个值，你能推出这两队最后的得分是多少吗？

2. 输入描述

输入的第一行是一个整数 n，表示测试案例的个数。接下来有 n 行，每行代表一个测试案例。每个测试案例由 s 和 d 两个数组成，都是非负整数，分别代表两个最后得分的和与差（的绝对值）。

3. 输出描述

对于每个测试案例，在一行上输出两个最终分数，大的在前面。如果不存在这样的分数，则在一行上输出"impossible"。再次提醒，足球分数是非负整数。

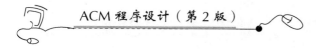

4. 输入样例

```
2
40 20
20 40
```

5. 输出样例

```
30 10
impossible
```

4.39 Champion of the Swordsmanship

4.39.1 链接地址

http://www.realoj.com/网上第 123 题

4.39.2 时空限制

Time Limit: 1000 ms　　Resident Memory Limit: 1024 KB　　Output Limit: 1024 B

4.39.3 题目内容

In Zhejiang University, there is a famous BBS named Freecity. Usually we call it 88.

Recently some students at the Humour board on 88 create a new game—Swordsmanship. Different from the common sword fights, this game can be held with three players playing together in a match. Only one player advances from the match while the other two are eliminated. Sometimes they also hold a two-player match if needed, but they always try to hold the tournament with as less matches as possible.

Input

The input contains several test cases. Each case is specified by one positive integer n ($0 < n < 1\,000\,000\,000$), indicating the number of players. Input is terminated by $n = 0$.

Output

For each test case, output a single line with the least number of matches needed to decide the champion.

Sample Input

```
3
4
0
```

Sample Output

```
1
2
```

4.39.4 题目来源

Zhejiang University Local Contest 2007, Preliminary (Author: XUE, Zaiyue)

4.39.5 解题思路

三个人可以进行一次比赛，两个人也可以举行一次比赛，问最少需要多少场次比赛。注意，不是问需要多少轮比赛。

如给定 3 个人，那么，共举行 1 次比赛即可。
如给定 4 个人，那么，共举行 2 次比赛即可。
如给定 6 个人，那么，共举行 3 次比赛即可。
如给定 12 个人，那么，共举行 6 次比赛即可。
下面以 12 个人为例，说明程序设计过程。

n 初值=12 人			
$d = n / 3$	$m = n \% 3$	新产生的 $n = m + d$	sum = sum + d
4	0	4	4
1	1	2	5

$n=1$ 时，不需再比赛。$n=2$ 时，还需再比一次。所以，sum = 6。

4.39.6 参考答案

```
#include <fstream>
#include <iostream>
#include <cmath>
using namespace std;
int main(int argc, char * argv[])
{
    ifstream cin("aaa.txt");
    int n,d,m,sum=0;
    while(cin>>n)
    {
        if(n==0)break;
        sum=0;
        while(1)
        {
            if(n==1)
            {
                cout<<sum<<endl;
                break;
            }
            if(n==2)
            {
                cout<<sum+1<<endl;
                break;
            }
```

```
            d=n/3;
            m=n%3;
            n=d+m;
            sum=sum+d;
        }
    }
    return 0;
}
```

4.39.7 汉语翻译

1. 题目

剑 术 冠 军

在浙江大学，有一个著名的 BBS，叫做 Freecity。我们通常把它称为 88。

最近，一些学生在 88 论坛的 Humour 栏目上创建了一个新的游戏——剑术。不同于常规的击剑比赛，这种游戏是三人同时进行一场比赛。一场比赛中，仅有一人晋级，而另外两人则被淘汰。有时，如果有必要的话，也举行两个人一场的比赛，但他们通常尽量让比赛的场数达到最少。

2. 输入描述

输入包含多个测试案例。每个测试案例是一个正整数 n（$0 < n < 1\,000\,000\,000$），表示参赛队员的人数。$n = 0$ 表示输入的终止。

3. 输出描述

对于每个测试案例，在一行上输出产生冠军需要的最少的比赛场次。

4. 输入样例

```
3
4
0
```

5. 输出样例

```
1
2
```

4.40　Doubles

4.40.1　链接地址

http://www.realoj.com/网上第 124 题

4.40.2 时空限制

Time Limit: 1000 ms   Resident Memory Limit: 1024 KB   Output Limit: 1024 B

4.40.3 题目内容

As part of an arithmetic competency program, your students will be given randomly generated lists of from 2 to 15 unique positive integers and asked to determine how many items in each list are twice some other item in the same list. You will need a program to help you with the grading. This program should be able to scan the lists and output the correct answer for each one. For example, given the list

1 4 3 2 9 7 18 22

your program should answer 3, as 2 is twice 1, 4 is twice 2, and 18 is twice 9.

Input

The input file will consist of one or more lists of numbers. There will be one list of numbers per line. Each list will contain from 2 to 15 unique positive integers. No integer will be larger than 99. Each line will be terminated with the integer 0, which is not considered part of the list. A line with the single number -1 will mark the end of the file. The example input below shows 3 separate lists. Some lists may not contain any doubles.

Output

The output will consist of one line per input list, containing a count of the items that are double some other item.

Sample Input

```
1 4 3 2 9 7 18 22 0
2 4 8 10 0
7 5 11 13 1 3 0
-1
```

Sample Output

```
3
2
0
```

4.40.4 题目来源

Mid-Central USA 2003

4.40.5 解题思路

本题要求找出一个列表中成两倍关系的元素的对数。

数据输入到集合中，自动会由小到大排序；再在集合中查找每个元素的两倍是否存在。由于集合是平衡二叉检索树（黑白树），因此查询速度是极快的。

4.40.6 参考答案

```
#include <fstream>
#include <iostream>
#include <set>
using namespace std;
int main(int argc, char * argv[])
{
    ifstream cin("aaa.txt");
    set<int>s;
    set<int>::iterator it,itt;
    int sum=0;
    int n;
    while(cin>>n)
    {
        if(n==-1)break;
        //读入一行数据
        if(n!=0)
        {
            s.insert(n);
        }
        //一行已读完了
        if(n==0)
        {
            //开始处理读入的这一行数据
            for(it=s.begin();it!=s.end();it++)
            {
                if(s.find(2*(*it))!=s.end())//查找到
                {
                    sum++;
                }
            }
            cout<<sum<<endl;
            //初始化各种变量与容器
            sum=0;
            s.clear();
        }
    }
    return 0;
}
```

4.40.7 汉语翻译

1. 题目

两　　倍

作为算术能力计划的一部分，你的学生将被给予一个随机产生的从 2 到 15 个不重复的正整数列表，要求说出这个表中有多少对数字是这个表中的其他数字的两倍。你需要一个程序来帮助你分级。这个程序会扫描这个列表并为每个测试案例输出正确的答案。比如，给定下面这个列表：

1 4 3 2 9 7 18 22

你的程序应当回答 3，由于 2 是 1 的两倍，4 是 2 的两倍，18 是 9 的两倍。

2. 输入描述

输入数据包含一列或多列数字。一行上有一列数字。每列包含 2～15 个非重复的正整数。所有的整数都不会大于 99。每行以 0 作为结束，不要把这个 0 作为列表的一部分。一行上只有一个整数-1 表示文件的结束。下面的输入样例中有 3 个独立的列表，有些列表没有包含任何两倍关系。

3. 输出描述

每个测试案例应当输出一行，打印出了这个测试案例中两倍关系的元素对的数目。

4. 输入样例

```
1 4 3 2 9 7 18 22 0
2 4 8 10 0
7 5 11 13 1 3 0
-1
```

5. 输出样例

```
3
2
0
```

4.41 File Searching

4.41.1 链接地址

http://www.realoj.com/网上第 125 题

4.41.2 时空限制

Time Limit: 1000 ms   Resident Memory Limit: 1024 KB   Output Limit: 1024 B

4.41.3 题目内容

Have you ever used file searching tools provided by an operating system? For example, in DOS, if you type "dir *.exe", the OS will list all executable files with extension "exe" in the current directory. These days, you are so mad with the crappy operating system you are using and you decide to write an OS of your own. Of course, you want to implement file searching functionality in your OS.

Input

The input contains several test cases. Consecutive test cases are separated by a blank line.

Each test case begins with an integer N ($1 \leq N \leq 100$), the number of files in the current directory. Then N lines follow, each line has one string consisting of lowercase letters ("a".. "z") and the dot (".") only, which is the name of a file. Then there is an integer M ($1 \leq M \leq 20$), the number of queries. M lines follow, each has one query string consisting of lowercase letters, the dot and the star ("*") character only. Note that the star character is the "universal matching character" which is used to represent zero or numbers of characters that are uncertain. In the beginning, you just want to write a simple version of file searching, so every string contains no more than 64 characters and there is one and only one star character in the query string.

Process to the End Of File (EOF).

Output

For each test case, generate one line for the results of each query. Separate file names in the result by a comma (",") and a blank (" ") character. The file names in the result of one query should be listed according to the order they appear in the input. If there is no matching file, output "FILE NOT FOUND" (without the quotation) instead.

Separate two consecutive test cases with a blank line, but **Do NOT** output an extra blank line after the last one.

Sample Input

```
4
command.com
msdos.sys
io.sys
config.sys
2
com*.com
*.sys

3
a.txt
b.txt
c.txt
1
*.doc
```

Sample Output

```
command.com
msdos.sys, io.sys, config.sys

FILE NOT FOUND
```

4.41.4 题目来源

Zhejiang University Local Contest 2007 (Author: ZHOU, Yuan)

4.41.5 解题思路

本题不容易，对编程基本功要求很高，且要求头脑清楚，否则，很难做出来。

查询串中只有一个"*"号，位置任意。它代表 0 个到多个任意字符。

只需将"*"号左边的字符和右边的字符取出，到文件名中的两头去比对就可。这种策略可以对付任何一种情况。

下面的测试案例比较完整（左边是输入数据，右边为输出结果）：

1	FILE NOT FOUND
beatea	
1	command.com
beat*tea	ms.dos.sys, io.sys, config.sys
	io.sys
5	
command.com	FILE NOT FOUND
commandcom9999999	
ms.dos.sys	FILE NOT FOUND
io.sys	
config.sys	a.b, a.aaa
3	
com*.com	io.sys
*.sys	
io.*	abcde
3	Press any key to continue
a.txt	
b.txt	
c.txt	
1	
*.doc	
1	
a.b	
1	
a.b*b	
3	
a.b	
aaa.aaa	

a.aaa 1 a.* 1 io.sys 1 i*o.sys 1 abcde 1 ab*cde	

4.41.6 参考答案

```
#include <fstream>
#include <iostream>
#include <string>
#include <vector>
#include <algorithm>
using namespace std;
int main(int argc, char * argv[])
{
    ifstream cin("aaa.txt");
    string s,ss,left,right;
    vector<string>v,end;
    int m,n;
    int i,j,k;
    int p,q;
    int c=0;
    while(cin>>m)
    {
        //案例数量
        c++;
        if(c>1)cout<<endl;
        //读入本案例所有文件名
        v.clear();
        for(i=0;i<m;i++)
        {
            cin>>s;
            v.push_back(s);
        }
        //读入要查询的文件名
        cin>>n;
        for(i=0;i<n;i++)
```

```cpp
{
    //读入一个要查询的文件名
    cin>>ss;
    //初始化 left 和 right
    left="";
    right="";
    //把 ss 按*号两边分出 left 和 right 两部分
    p=ss.find("*");
    //区分出 left
    for(j=0;j<p;j++)
        left+=ss[j];
    //区分出 right
    for(j=p+1;j<ss.length();j++)
        right+=ss[j];
    //清空查询结果容器
    end.clear();
    //开始查找
    for(j=0;j<v.size();j++)
    {
        //文件名长度不能小于查询串*号左边加右边的长度
        if(v[j].size()<(left.size()+right.size()))
            continue;
        //比较查询串*号左边串
        if(left.size()!=0)
        {
            if(v[j].find(left)!=0)continue;
        }
        //比较查询串*号右边串
        if(right.size()!=0)
        {
            //反转
            reverse(right.begin(),right.end());
            reverse(v[j].begin(),v[j].end());
            if(v[j].find(right)!=0)
            {
                reverse(right.begin(),right.end());
                reverse(v[j].begin(),v[j].end());
                continue;
            }
            //重新反转来
            reverse(right.begin(),right.end());
            reverse(v[j].begin(),v[j].end());
        }
        //符合条件的放入向量中
        end.push_back(v[j]);
    }
    //输出一次查询结果
    for(k=0;k<end.size();k++)
    {
        cout<<end[k];
```

```cpp
                    if(k!=end.size()-1)
                        cout<<", ";
                    else
                        cout<<endl;
            }
            //如果没有查出来，就打印"FILE NOT FOUND"
            if(end.size()==0)
                cout<<"FILE NOT FOUND"<<endl;
        }
    }
    return 0;
}
```

4.41.7 汉语翻译

1. 题目

查 找 文 件

你曾经使用过由操作系统提供的文件查找工具吗？例如，在 DOS 操作系统中，如果你输入"dir *.exe"，那么操作系统就会列出当前目录中扩展名为"exe"的所有可执行文件。这些天来，你对你正在使用的讨厌的操作系统发疯了，你决定编写一个自己的操作系统。当然，你要在你的操作系统中编写文件查找功能。

2. 输入描述

输入数据包含多个测试案例。连续的测试案例之间用一行空行隔开。

每个测试案例的开头是一个整数 N（$1 \leqslant N \leqslant 100$），表示当前目录中文件名的数量。然后 N 行，每行上是一个仅由小写字母（"a".."z"）和点（"."）组成的字符串，它是一个文件的文件名。然后，是一下整数 M（$1 \leqslant M \leqslant 20$），表示查询的数量。接下去是 M 行，每行是一个由小写字母、点号和星号（"*"）组成。注意，星号是一个全部匹配字符，代表 0 个或任意多个任意字符。开始，你只想写一个简单的文件查找程序，所以，每个字符串不超过 64 个字符，有且仅有一个星号在查询字符串中。

一直处理到文件尾（EOF）。

3. 输出描述

对于每个测试案例，把查询结果打印在一行上，文件名间用逗号（','）和一个空格字符（' '）隔开。输出的文件名与它们在文件列表中的先后顺序一致。如果找不到文件，则输出"FILE NOT FOUND"（不要引号）代替。

两个测试案例间用一行空行隔开，但不要在最后一个案例的后边再输出一行空行。

4. 输入样例

```
4
command.com
msdos.sys
io.sys
```

```
config.sys
2
com*.com
*.sys

3
a.txt
b.txt
c.txt
1
*.doc
```

5. 输出样例

```
command.com
msdos.sys, io.sys, config.sys

FILE NOT FOUND
```

4.42 Old Bill

4.42.1 链接地址

http://www.realoj.com/网上第 126 题

4.42.2 时空限制

Time Limit: 1000 ms   Resident Memory Limit: 1024 KB   Output Limit: 1024 B

4.42.3 题目内容

Among grandfather's papers a bill was found:

$$72 \text{ turkeys } \$\_679\_$$

The first and the last digits of the number that obviously represented the total price of those turkeys are replaced here by blanks (denoted _), for they are faded and are now illegible. What are the two faded digits and what was the price of one turkey?

We want to write a program that solves a general version of the above problem:

$$N \text{ turkeys } \$\_XYZ\_$$

The total number of turkeys, N, is between 1 and 99, including both. The total price originally consisted of five digits, but we can see only the three digits in the middle. We assume that the first digit is nonzero, that the price of one turkey is an integer number of dollars, and that all the turkeys cost the same price.

Given N, X, Y, and Z, write a program that guesses the two faded digits and the original price. In case that there is more than one candidate for the original price, the output should be the most expensive one. That is, the program is to report the two faded digits and the maximum price per turkey for the turkeys.

Input

The input consists of T test cases. The number of test cases (T) is given on the first line of the input file. The first line of each test case contains an integer N ($0 < N < 100$), which represents the number of turkeys. In the following line, there are the three decimal digits X, Y, and Z, separated by a space, of the original price $\_XYZ\_$.

Output

For each test case, your program has to do the following. For a test case, there may be more than one candidate for the original price or there is none. In the latter case your program is to report 0. Otherwise, if there is more than one candidate for the original price, the program is to report the two faded digits and the maximum price per turkey for the turkeys. The following shows sample input and output for three test cases.

Sample Input

```
3
72
6 7 9
5
2 3 7
78
0 0 5
```

Sample Output

```
3 2 511
9 5 18475
0
```

4.42.4　题目来源

Asia 2003, Seoul (South Korea)

4.42.5　解题思路

本题的主要难点是在读题上，如果读懂了题意，还是比较好实现的。

本题是穷举题，就是把所有符合要求的方案都枚举出来。要求的条件是：首位不能为0；另外，兑换比例要为整数。打印出价格最大的那组数据，如果找不到这样的组合，那么打印出一个0。

4.42.6　参考答案

```cpp
#include <fstream>
#include <iostream>
#include <vector>
```

```cpp
#include <algorithm>
using namespace std;
//结构体，用来保存输出数据的结构
struct Info{
    int first;
    int last;
    int average;
};
//自定义比较函数，按兑换比例由大到小排序
bool comp(const Info &a,const Info &b)
{
    if(a.average!=b.average)return a.average>b.average;
}
int main(int argc, char * argv[])
{
    ifstream cin("aaa.txt");
    Info info;
    vector<Info>v;
    int i,j;
    int n;
    int x,y,z,m;
    cin>>n;//不理会第1个数据
    while(cin>>n)
    {
        cin>>x>>y>>z;
        m=x*1000 + 100*y + 10*z;
        //清除向量
        v.clear();
        //计算需要的数据
        for(i=1;i<10;i++)
        {
            for(j=0;j<10;j++)
            {
                if((i*10000+m+j)%n==0)
                {
                    info.first=i;
                    info.last=j;
                    info.average=(i*10000+m+j)/n;
                    v.push_back(info);
                }
            }
        }
        //按兑换比例数由大到小排序
        sort(v.begin(),v.end(),comp);
        //输出数据
        if(v.size()==0)
            cout<<"0"<<endl;
        else
            cout<<v[0].first<<" "<<v[0].last<<" "<<v[0].average<<endl;
    }
```

```
        return 0;
    }
```

4.42.7 汉语翻译

1. 题目

旧　账　单

在祖父的文件里，发现了一张账单：

72 turkeys $_679_

表示这些火鸡总价格的数字的第一位和最后一位用"_"线代替了，因为这两位褪色了，难以辨认。这两位褪色的数字是什么呢？一只火鸡的价格是多少呢？

我们想写一个程序来计算上面的问题，样式是：

N turkeys $_XYZ_

N 表示火鸡的总数目，在 1～99 的一个闭区间中。总价格包括 5 位数，但我们只能看到中间的 3 位数。假定第一位是非 0，且一只火鸡的价格是一个整数美元，另外，所有火鸡的价格是相同的。

给定 N，X，Y 和 Z，写一个程序猜出它褪色的两位数字和它的原始价格。这里可能会出现多种原始价格，你只要输出最贵的一个价格。也就是说，程序应当报告两位褪色的数字和每只火鸡最高的价格。

2. 输入描述

输入数据包含 T 个测试案例。测试案例 T 在输入文件的第一行上。每个测试案例的第一行是一个整数 N（0 < N < 100），代表火鸡的总数。在接下去的一行中，有 3 个十进制数字 X，Y 和 Z，它们之间用一个空格隔开，它们代表这些火鸡的原始价格$_XYZ_。

3. 输出描述

对于每个测试案例，你的程序要这样做：对于每个测试案例，可能不止一种价格方案，也可能没有一种价格方案。对于后一种情况，你的程序要报告 0；否则，如果不止一种价格方案，程序报告这两位褪色的数字，这两个数字是最贵的那个价格中的。下面举了 3 个测试案例，显示了如何进行数据的输入和输出。

4. 输入样例

```
3
72
6 7 9
5
2 3 7
78
0030 0 5
```

5. 输出样例

```
3 2 511
9 5 18475
0
```

4.43 Divisor Summation

4.43.1 链接地址

http://www.realoj.com/网上第 127 题

4.43.2 时空限制

Time Limit: 1000 ms   Resident Memory Limit: 1024 KB   Output Limit: 1024 B

4.43.3 题目内容

Give a natural number n ($1 \leq n \leq 500\,000$), please tell the summation of all its proper divisors.

Definition: A proper divisor of a natural number is the divisor that is strictly less than the number.

e.g. number 20 has 5 proper divisors: 1, 2, 4, 5, 10, and the divisor summation is: $1 + 2 + 4 + 5 + 10 = 22$.

Input

An integer stating the number of test cases, and that many lines follow each containing one integer between 1 and 500 000.

Output

One integer each line: the divisor summation of the integer given respectively.

Sample Input

```
3
2
10
20
```

Sample Output

```
1
8
22
```

4.43.4 题目来源

ZOJ Monthly, March 2004 (Author: Neal Zane)

4.43.5 解题思路

本题题意比较明朗，如果采用穷举法，会出现超时错误。应当先建立一个[2,500 000]的表，最后直接查表即可。

本题采用 cin 和 cout 输入输出，会严重超时，而采用 scanf 和 printf 输入输出就不会超时。另外，要注意，如果是 1，则输出 0。因为 1 没有严格意义上的除数（proper divisors）。

4.43.6 参考答案

```cpp
#include <fstream>
#include <iostream>
#include <cmath>
using namespace std;
int m[500000];
int main(int argc, char * argv[])
{
    ifstream cin("aaa.txt");
    int i,j;
    int n;
    int d;
    d=sqrt(500000.0);
    //1 是每个数的约数
    m[1]=0;//1 没有除 1 外的约数
    for(i=2;i<=500000;i++)
        m[i]=1;
    //开始统计每个数的约数和
    for(i=2;i<=d;i++)
    {
        m[i*i]+=i;
        for(j=i+1;j<=(500000/i);j++)
        {
            m[i*j]+=(i+j);
        }
    }
    cin>>n;//不理会第一个数据
    //采用 cin 输入，会严重超时
    //while(cin>>n)
    while(scanf("%d",&n)!=EOF)
    {
        //采用 cout 输出，会严重超时
        //cout<<m[n]<<endl;
        printf("%d\n",m[n]);
    }
    return 0;
}
```

4.43.7 汉语翻译

1. 题目

除 数 的 和

给定一个自然数 n（$1 \leqslant n \leqslant 500\,000$），请说出它的所有严格意义上的除数的和。

定义：一个自然数的严格意义上的除数是比它本身小的除数。

比如，20 有 5 个严格意义上的除数：1，2，4，5，10，所以，除数的和是 $1 + 2 + 4 + 5 + 10 = 22$。

2. 输入描述

第一个整数是表示测试案例的个数，然后是多行，每行上有一个 1 到 500 000 的整数。

3. 输出描述

每行输出一个整数：表示每个测试案例的除数和。

4. 输入样例

```
3
2
10
20
```

5. 输出样例

```
1
8
22
```

4.44 Easier Done Than Said?

4.44.1 链接地址

http://www.realoj.com/网上第 128 题

4.44.2 时空限制

Time Limit: 1000 ms   Resident Memory Limit: 1024 KB   Output Limit: 1024 B

4.44.3 题目内容

Password security is a tricky thing. Users prefer simple passwords that are easy to remember (like buddy), but such passwords are often insecure. Some sites use random computer-generated passwords (like xvtpzyo), but users have a hard time remembering them and sometimes leave them written on notes stuck to their computer. One potential solution is to generate "pronounceable" passwords that are relatively secure but still easy to remember.

FnordCom is developing such a password generator. You work in the quality control department, and it's your job to test the generator and make sure that the passwords are acceptable. To be acceptable, a password must satisfy these three rules:

(1) It must contain at least one vowel.

(2) It cannot contain three consecutive vowels or three consecutive consonants.

(3) It cannot contain two consecutive occurrences of the same letter, except for "ee" or "oo".

(For the purposes of this problem, the vowels are "a", "e", "i", "o", and "u"; all other letters are consonants.) Note that these rules are not perfect; there are many common/pronounceable words that are not acceptable.

Input

The input consists of one or more potential passwords, one per line, followed by a line containing only the word "end" that signals the end of the file. Each password is at least one and at most twenty letters long and consists only of lowercase letters.

Output

For each password, output whether or not it is acceptable, using the precise format shown in the example.

Sample Input

```
a
tv
ptoui
bontres
zoggax
wiinq
eep
houctuh
end
```

Sample Output

```
<a> is acceptable.
<tv> is not acceptable.
<ptoui> is not acceptable.
<bontres> is not acceptable.
<zoggax> is not acceptable.
<wiinq> is not acceptable.
<eep> is acceptable.
<houctuh> is acceptable.
```

4.44.4 题目来源

Mid-Central USA 2000

4.44.5 解题思路

本题是字符串扫描程序，要求有很高的编程功底，题目也很好看懂。一个密码，如果符合要求，则输出可接受，否则输出不可接受。

一个密码只有都满足了以下三种情况才是可以接受的：

（1）字符串中至少包含一个元音字母，a, e, i, o 或 u。

（2）字符串中不能包含三个连续元音或三个连续辅音。

（3）不能包含两个相同的连续字母，但 ee 和 oo 除外。

4.44.6 参考答案

```cpp
#include <fstream>
#include <iostream>
#include <string>
using namespace std;
int main(int argc, char * argv[])
{
    ifstream cin("aaa.txt");
    string s;
    int a,b;
    int i,t;
    while(cin>>s)
    {
        if(s=="end")break;
        //条件1：必须含有一个元音,a,e i,o,u是元音
        if(s.find("a")>s.size() && s.find("e")>s.size() && s.find("i")
            >s.size() && s.find("o")>s.size() && s.find("u")>s.size())
        {
            cout<<"<"<<s<<">"<<" is not acceptable."<<endl;
            continue;
        }
        //条件2：不能包含3个连续元音或三个连续辅音
        a=0;//连续元音个数
        b=0;//连续辅音个数
        for(i=0;i<s.size();i++)
        {
            //判断是元音还是辅音字母
            if(s[i]=='a' || s[i]=='e' || s[i]=='i' || s[i]=='o' || s[i]=='u')
            {
                if(i==0)a=1;
                else
                {
                    if(s[i-1]=='a' || s[i-1]=='e' || s[i-1]=='i' || s[i-1]
                        =='o' || s[i-1]=='u')
                    {
                        a++;
                        b=0;
```

```
                    }
                    else
                    {
                        a=1;
                        b=0;
                    }
                }
            }
            else
            {
                if(i==0)b=1;
                else
                {
                    if(s[i-1]!='a' || s[i-1]!='e' || s[i-1]!='i' || s[i-1]!='o' || s[i-1]!='u')
                    {
                        b++;
                        a=0;
                    }
                    else
                    {
                        b=1;
                        a=0;
                    }
                }
            }
            //出现三个连续的,就终止
            if(a==3 || b==3)
            {
                cout<<"<"<<s<<">"<<" is not acceptable."<<endl;
                goto RL;
            }
        }
        //条件3:不能包含两个连续字母,但ee和oo除外
        t=0;
        for(i=0;i<s.size();i++)
        {
            if(i==0)t=1;
            else if(s[i]=='e' || s[i]=='o')
                t=0;
            else
            {
                if(s[i]==s[i-1])t++;
                else t=1;
            }
            if(t==2)
            {
                cout<<"<"<<s<<">"<<" is not acceptable."<<endl;
                goto RL;
            }
```

```
        }
        cout<<"<"<<s<<">"<<" is acceptable."<<endl;
RL:
        continue;
    }
    return 0;
}
```

4.44.7 汉语翻译

1. 题目

做比说还容易？

密码安全是一件费脑子的事情。人们喜欢使用简单的密码，这样好记（如 buddy），但这样的密码通常不太安全。但一些网站计算机随机生成的密码（如 xvtpxyo），却太难记了，有时候只得把它们写在小纸条上并贴在自己的电脑上。一个潜在的解决方案是产生"可发音的"的密码，这种密码相对安全且易记忆。

FnordCom 正在开发这样的密码生成器。你工作在质检部门，测试这个密码生成器且确定密码的可用性是你的工作。可按受的密码，必须满足以下三条规则：

（1）它必须包含至少一个元音字母。
（2）它不能包含三个连续的元音字母或三个连续的辅音字母。
（3）它不能包含两个连续的相同的字母，除了"ee"或"oo"之外。
（对于本题，元音字母是"a"，"e"，"i"，"o"和"u"，其他字母都是辅音。）
注意，这些规则并不是十分完美；有很多常用的可发音的单词都不能被接受。

2. 输入描述

输入数据包含一个或多个潜在密码，一行一个，"end"表示输入的结束。每个密码的长度是1到20，且只由小写字母组成。

3. 输出描述

对于每个密码，输出它是否可以被接受，使用样例中的精确的格式。

4. 输入样例

```
a
tv
ptoui
bontres
zoggax
wiinq
eep
houctuh
end
```

5. 输出样例

```
<a> is acceptable.
<tv> is not acceptable.
<ptoui> is not acceptable.
<bontres> is not acceptable.
<zoggax> is not acceptable.
<wiinq> is not acceptable.
<eep> is acceptable.
<houctuh> is acceptable.
```

4.45　Let the Balloon Rise

4.45.1　链接地址

http://www.realoj.com/ 网上第 129 题

4.45.2　时空限制

Time Limit: 1000 ms　　Resident Memory Limit: 1024 KB　　Output Limit: 1024 B

4.45.3　题目内容

Contest time again! How excited it is to see balloons floating around. But to tell you a secret, the judges' favorite time is guessing the most popular problem. When the contest is over, they will count the balloons of each color and find the result.

This year, they decide to leave this lovely job to you.

Input

Input contains multiple test cases. Each test case starts with a number N ($0 < N < 1000$)—the total number of balloons distributed. The next N lines contain one color each. The color of a balloon is a string of up to 15 lower-case letters.

A test case with $N = 0$ terminates the input and this test case is not to be processed.

Output

For each case, print the color of balloon for the most popular problem on a single line. It is guaranteed that there is a unique solution for each test case.

Sample Input

```
5
green
red
blue
red
red
```

```
3
pink
orange
pink
0
```

Sample Output

```
red
pink
```

4.45.4 题目来源

Zhejiang Provincial Programming Contest 2004 (Author: WU, Jiazhi)

4.45.5 解题思路

本题实际上就是统计每种颜色出现的数量，不好的算法可能使得编程实现起来比较困难且可能导致超时。

使用映照容器比较好实现，速度也快。键值是气球颜色的单词，映照数据则是该种颜色的气球出现的次数。

先把数据统计到映照容器中后，再搜索出映照数据中大的那个元素，最后把该元素的键值打印出来就好了。

4.45.6 参考答案

```
#include <fstream>
#include <iostream>
#include <string>
#include <map>
using namespace std;
int main(int argc, char * argv[])
{
    ifstream cin("aaa.txt");
    map<string,int>m;
    int n;
    int i;
    string s;
    map<string,int>::iterator it,it2;
    while(cin>>n)
    {
        if(n==0)break;
        m.clear();
        for(i=0;i<n;i++)
        {
            cin>>s;
            //在映照容器中查找键值
            if(m.find(s)!=m.end())
                m[s]=m[s]+1;
```

```
            else
                m[s]=1;
        }
        it2=m.begin();
        for(it=m.begin();it!=m.end();it++)
        {
            //注意，映照容器中，元素的键值为 it->first,
            //元素映照值为 it->second
            if(it2->second<it->second)
                it2=it;
        }
        cout<<it2->first<<endl;
    }
    return 0;
}
```

4.45.7 汉语翻译

1. 题目

让气球升起来

比赛又要开始了！看到气球四处升起是一件多么令人激动的事情啊。但告诉你一个秘密，裁判最爱做的事情就是猜哪道题最热门。当比赛一结束，他们就要数出每个气球的颜色从而得出哪道题最热门的结论。

今年，这项有趣的工作就留给你去做。

2. 输入描述

输入数据包含多组测试案例。每组测试案例由 N（$0 < N < 1\,000$）打头，N 表示分发的气球的总数目。接下去的 N 行每行包含一种颜色。气球的颜色是一个多达 15 个小写字母的单词。

一个测试案例的 $N = 0$ 表示输入的结束，你不要去处理这个测试案例。

3. 输出描述

对于每个测试案例，把数目最多的那种颜色打印在单独一行上。每个测试案例都仅有一种颜色是最多的。

4. 输入样例

```
5
green
red
blue
red
red
3
pink
orange
```

```
pink
0
```

5. 输出样例

```
red
pink
```

4.46 The Hardest Problem Ever

4.46.1 链接地址

http://www.realoj.com/网上第 130 题

4.46.2 时空限制

Time Limit: 1000 ms   Resident Memory Limit: 1024 KB   Output Limit: 1024 B

4.46.3 题目内容

Introduction

Julius Caesar lived in a time of danger and intrigue. The hardest situation Caesar ever faced was keeping himself alive. In order for him to survive, he decided to create one of the first ciphers. This cipher was so incredibly sound, that no one could figure it out without knowing how it worked.

You are a sub captain of Caesar's army. It is your job to decipher the messages sent by Caesar and provide to your general. The code is simple. For each letter in a plaintext message, you shift it five places to the right to create the secure message (i.e., if the letter is "A", the cipher text would be "F"). Since you are creating plain text out of Caesar's messages, you will do the opposite:

Cipher text
A B C D E F G H I J K L M N O P Q R S T U V W X Y Z
Plain text
V W X Y Z A B C D E F G H I J K L M N O P Q R S T U

Only letters are shifted in this cipher. Any non-alphabetical character should remain the same, and all alphabetical characters will be upper case.

Input

Input to this problem will consist of a (non-empty) series of up to 100 data sets. Each data set will be formatted according to the following description, and there will be no blank lines separating data sets. All characters will be uppercase.

A single data set has 3 components:

(1) Start line—A single line, "START"

(2) Cipher message—A single line containing from one to two hundred characters, inclusive, comprising a single message from Caesar

(3) End line—A single line, "END"

Following the final data set will be a single line, "ENDOFINPUT".

Output

For each data set, there will be exactly one line of output. This is the original message by Caesar.

Sample Input

```
START
NS BFW, JAJSYX TK NRUTWYFSHJ FWJ YMJ WJXZQY TK YWNANFQ HFZXJX
END
START
N BTZQI WFYMJW GJ KNWXY NS F QNYYQJ NGJWNFS ANQQFLJ YMFS XJHTSI NS WTRJ
END
START
IFSLJW PSTBX KZQQ BJQQ YMFY HFJXFW NX RTWJ IFSLJWTZX YMFS MJ
END
ENDOFINPUT
```

Sample Output

```
IN WAR, EVENTS OF IMPORTANCE ARE THE RESULT OF TRIVIAL CAUSES
I WOULD RATHER BE FIRST IN A LITTLE IBERIAN VILLAGE THAN SECOND IN ROME
DANGER KNOWS FULL WELL THAT CAESAR IS MORE DANGEROUS THAN HE
```

4.46.4 题目来源

South Central USA 2002

4.46.5 解题思路

本题是密码破译题，只处理字母。由于密文与明文中的字母是一一对应的，所以，本题采用 map 映照容器速度最快，也算是比较经典的 map 应用题。

另外，采用 cout 输出，不必考虑那些 printf 输出中的转义字符，如 "\n" 等。

4.46.6 参考答案

```
#include <fstream>
#include <iostream>
#include <map>
#include <string>
using namespace std;
int main(int argc, char * argv[])
{
    ifstream cin("aaa.txt");
    string s;
```

```cpp
char ss[200];
map<char,char>m;
m['A']='V';
m['B']='W';
m['C']='X';
m['D']='Y';
m['E']='Z';
m['F']='A';
m['G']='B';
m['H']='C';
m['I']='D';
m['J']='E';
m['K']='F';
m['L']='G';
m['M']='H';
m['N']='I';
m['O']='J';
m['P']='K';
m['Q']='L';
m['R']='M';
m['S']='N';
m['T']='O';
m['U']='P';
m['V']='Q';
m['W']='R';
m['X']='S';
m['Y']='T';
m['Z']='U';
int i;
while(cin.getline(ss,200))
{
    s=ss;
    if(s=="START")continue;
    else if(s=="END")continue;
    else if(s=="ENDOFINPUT")break;
    else
    {
        for(i=0;i<s.size();i++)
        {
            if(s[i]>='A' && s[i]<='Z')
            {
                cout<<m[s[i]];
            }
            else
                cout<<s[i];
        }
        cout<<endl;
    }
}
```

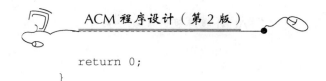

```
    return 0;
}
```

4.46.7 汉语翻译

1. 题目

曾经是最难的问题

介绍

凯撒大帝生活在一个充满危险和阴谋的时期。凯撒面临的最危险的情况就是保全性命。为了生存下去，他决定创建第一套密码。这套密码听起来如此难以置信，如果不知道它的工作原理，没有人能够揣摩出明文。

你是凯撒的军队的一个分队长。你的工作就是破译凯撒的消息然后提供给你的将军。代码很简单。对于明文中的每个字母，你把它向字母表里向右移动 5 个位置就创建了它的密文（如，如果是字母是"A"，密文就是"F"）。既然你是把凯撒的密文破译为明文，那么你将是反过来做。

密文：

A B C D E F G H I J K L M N O P Q R S T U V W X Y Z

明文：

V W X Y Z A B C D E F G H I J K L M N O P Q R S T U

在密文中，只需移动字母。任何非英文字母都应该保留原样，所有的字母都是大写。

2. 输入描述

本题的输入数据由多达 100 个非空数据集组成。每个数据集的格式是按照以下描述的格式，数据集间没有空行。所有的字符都是大写字母。

一个数据集包含三个部分：

（1）起始行——是一行字符串"START"。

（2）密文信息——一行，包含 100~200 个字符，是按凯撒密码产生的一行信息。

（3）结束行——是一个字符串"END"。

"ENDOFINPUT"在单独一行上，表示所有数据集的输入结束。

3. 输出描述

对于每个数据集，只输出一行，即通过凯撒密码破译的明文。

4. 输入样例

```
START
NS BFW, JAJSYX TK NRUTWYFSHJ FWJ YMJ WJXZQY TK YWNANFQ HFZXJX
END
START
N BTZQI WFYMJW GJ KNWXY NS F QNYYQJ NGJWNFS ANQQFLJ YMFS XJHTSI NS WTRJ
```

```
END
START
IFSLJW PSTBX KZQQ BJQQ YMFY HFJXFW NX RTWJ IFSLJWTZX YMFS MJ
END
ENDOFINPUT
```

5. 输出样例

```
IN WAR, EVENTS OF IMPORTANCE ARE THE RESULT OF TRIVIAL CAUSES
I WOULD RATHER BE FIRST IN A LITTLE IBERIAN VILLAGE THAN SECOND IN ROME
DANGER KNOWS FULL WELL THAT CAESAR IS MORE DANGEROUS THAN HE
```

4.47　Fibonacci Again

4.47.1　链接地址

http://www.realoj.com/网上第 131 题

4.47.2　时空限制

Time Limit: 1000 ms　　Resident Memory Limit: 1024 KB　　Output Limit: 1024 B

4.47.3　题目内容

There are another kind of Fibonacci numbers: $F(0) = 7$, $F(1) = 11$, $F(n) = F(n-1) + F(n-2)$ ($n \geq 2$).

Input

Input consists of a sequence of lines, each containing an integer n. ($n < 1\,000\,000$)

Output

Print the word "yes" if 3 divide evenly into $F(n)$.

Print the word "no" if not.

Sample Input

```
0
1
2
3
4
5
```

Sample Output

```
no
no
yes
no
```

no
no

4.47.4 题目来源

ZOJ Monthly, December 2003 (Author: Leojay)

4.47.5 解题思路

只要能被 3 整除就输出 "yes"，否则，就输出 "no"。

先把 1 000 000 项都计算出来，放在向量中，再进行查询，是比较好的策略。但项数可达到 1 000 000 项，由于会超过无符号整型（32 位）的表达范围，所以，参考答案中第（1）种解法是错误的（Wrong Answer）。

对每位先对 3 取余（实际上是对各位减去 3 的位数），建立一个有 1 000 000 个元素的表，问题就解决了。参考答案中第（2）种解法就是采用这一思想。

最后，再试试另一种方法，现在，把前 20 项打印出来：

no
no
yes
no
no
no
yes
no
no
no
yes
no
no
no
yes
no
no
no
yes
no
no

可以发现，每隔 3 个 "no" 就有一个 "yes"，参考答案中第（3）种解法就是采用这种思路，也可以通过，还快速。

从这道题我们可以得到一个启示，ACM 竞赛中试题解法是很多的，我们的目的是能第一时间提交且通过。

4.47.6 参考答案

（1）错误的解法。

```cpp
#include <fstream>
#include <iostream>
#include <vector>
using namespace std;
vector<unsigned int> v;
int main(int argc, char * argv[])
{
    ifstream cin("aaa.txt");
    unsigned int n1,n2,t;
    unsigned int n;
    int i;
    n1=7;
    n2=11;
    v.push_back(7);
    v.push_back(11);
    for(i=2;i<=1000000;i++)
    {
        v.push_back(n1+n2);
        t=n1+n2;
        n1=n2;
        n2=t;
    }
    while(cin>>n)
    {
        if(v[n]%3==0)
            cout<<"yes"<<endl;
        else
            cout<<"no"<<endl;
    }
    return 0;
}
```

（2）正确的解法。

```cpp
#include <fstream>
#include <iostream>
#include <vector>
using namespace std;
vector<unsigned int> v;
int main(int argc, char * argv[])
{
    ifstream cin("aaa.txt");
    unsigned int n1,n2,t;
    unsigned int n;
    int i;
```

```
            n1=7%3;
            n2=11%3;
            v.push_back(n1);
            v.push_back(n2);
            for(i=2;i<=1000000;i++)
            {
                t=(n1+n2)%3;
                v.push_back(t);
                n1=n2;
                n2=t;
            }
            while(cin>>n)
            {
                if(v[n]%3==0)
                    cout<<"yes"<<endl;
                else
                    cout<<"no"<<endl;
            }
            return 0;
        }
```

（3）正确的解法。

```
#include <fstream>
#include <iostream>
using namespace std;
int main(int argc, char * argv[])
{
    ifstream cin("aaa.txt");
    int n;
    while(cin>>n)
    {
        if(n%4==2)
            cout<<"yes"<<endl;
        else
            cout<<"no"<<endl;
    }
    return 0;
}
```

4.47.7 汉语翻译

1. 题目

又是斐波纳契数列

有另一种斐波纳契数列：$F(0) = 7$，$F(1) = 11$，$F(n) = F(n-1) + F(n-2)$（$n \geq 2$）。

2. 输入描述

输入由多行组成，每一行上是一个整数 n（$n < 1\,000\,000$）。

3. 输出描述

如果 $F(n)$ 能被 3 整除，那么打印一行"yes"；否则，打印一行"no"。

4. 输入样例

```
0
1
2
3
4
5
```

5. 输出样例

```
no
no
yes
no
no
no
```

4.48 Excuses, Excuses!

4.48.1 链接地址

http://www.realoj.com/网上第 132 题

4.48.2 时空限制

Time Limit: 1000 ms　　Resident Memory Limit: 1024 KB　　Output Limit: 1024 B

4.48.3 题目内容

Judge Ito is having a problem with people subpoenaed for jury duty giving rather lame excuses in order to avoid serving. In order to reduce the amount of time required listening to goofy excuses, Judge Ito has asked that you write a program that will search for a list of keywords in a list of excuses identifying lame excuses. Keywords can be matched in an excuse regardless of case.

Input

Input to your program will consist of multiple sets of data.

Line 1 of each set will contain exactly two integers. The first number ($1 \leq K \leq 20$) defines the number of keywords to be used in the search. The second number ($1 \leq E \leq 20$) defines the number of excuses in the set to be searched.

Lines 2 through $K+1$ each contain exactly one keyword.

Lines K+2 through K+1+E each contain exactly one excuse.

All keywords in the keyword list will contain only contiguous lower case alphabetic characters of length L (1≤L≤20) and will occupy columns 1 through L in the input line.

All excuses can contain any upper or lower case alphanumeric character, a space, or any of the following punctuation marks [SPMamp".,!?&] not including the square brackets and will not exceed 70 characters in length.

Excuses will contain at least 1 non-space character.

Output

For each input set, you are to print the worst excuse(s) from the list.

The worst excuse(s) is/are defined as the excuse(s) which contains the largest number of incidences of keywords.

If a keyword occurs more than once in an excuse, each occurrence is considered a separate incidence.

A keyword "occurs" in an excuse if and only if it exists in the string in contiguous form and is delimited by the beginning or end of the line or any non-alphabetic character or a space.

For each set of input, you are to print a single line with the number of the set immediately after the string "Excuse Set #" (See the Sample Output). The following line(s) is/are to contain the worst excuse(s) one per line exactly as read in. If there is more than one worst excuse, you may print them in any order.

After each set of output, you should print a blank line.

Sample Input

```
5 3
dog
ate
homework
canary
died
My dog ate my homework.
Can you believe my dog died after eating my canary... AND MY HOMEWORK?
This excuse is so good that it contain 0 keywords.
6 5
superhighway
crazy
thermonuclear
bedroom
war
building
I am having a superhighway built in my bedroom.
I am actually crazy.
1234567890.....,,,,,0987654321?????!!!!!
There was a thermonuclear war!
I ate my dog, my canary, and my homework ... note outdated keywords?
```

Sample Output

```
Excuse Set #1
Can you believe my dog died after eating my canary... AND MY HOMEWORK?

Excuse Set #2
I am having a superhighway built in my bedroom.
There was a thermonuclear war!
```

4.48.4 题目来源

South Central USA 1996

4.48.5 解题思路

本题需要相当高的编程基本功，考查的关键点是从语句行中查找关键词最多的一行，打印出来。

注意，一个关键词可以重复出现多次，都算是不同的情况，都需要计算。

另外，如果一个关键词在开头出现，那么，它的后面不能为字母；如果一个关键词在其他位置出现，那么，它的前面和后面一个位置都不能是字母。

一行的最后，是以标点符号结束的。

在输入时注意，cin 单个输入后，再采用 cin.getline()输入一行字符串时，第一次读出的是空行（即只含一个回车符号），需要跳过这个空行。主要原因是 cin 会忽略回车符和空格等。

4.48.6 参考答案

```cpp
#include <fstream>
#include <iostream>
#include <string>
#include <vector>
#include <algorithm>
using namespace std;
//结构体，用来保存语句行及其包含的关键词个数
struct Info{
    //字符行
    string s;
    //本字符行包含了关键词个数
    int n;
};
//自定义比较函数，按包含关键词的多少，由大到小排列
bool comp(const Info &a,const Info &b)
{
    if(a.n!=b.n)return a.n>b.n;
    else
        return a.n>b.n;
}
```

```cpp
int main(int argc, char * argv[])
{
    ifstream cin("aaa.txt");
    Info info;
    int m,n;
    char ss[71];
    string s,t;
    vector<string>v;//保存关键词
    vector<Info>vv;//保存语句行结构体
    int line=0;
    int i,j,k;
    unsigned int p;
    while(cin>>m>>n)
    {
        v.clear();
        vv.clear();
        line++;//案例号
        //读入关键词
        for(i=0;i<m;i++)
        {
            cin>>ss;
            s=ss;
            v.push_back(s);
        }
        //忽略空行,这句至关重要
        //因为cin读入单项后,仍然有一个回车符留在本行
        //这样,直接转入cin.getline()读入一整行,第一次读入的是个空行
        cin.getline(ss,71);
        //读入语句行
        for(i=0;i<n;i++)
        {
            cin.getline(ss,71);
            s=ss;
            //为了实现不区分大小写比较,
            //先把本行都变成小写后再比较
            for(j=0;j<s.size();j++)
            {
                //如果是大写字母则转化为小写字母
                if(s[j]>=65 && s[j]<=90)
                    s[j]+=32;
            }
            info.n=0;
            //搜索本行包含多少个关键词
            for(j=0;j<v.size();j++)
            {
                for(k=0;k<s.size()-v[j].size();k++)
                {
                    //取出一个子串
                    t="";
                    for(p=k;p<k+v[j].size();p++)
```

```
                {
                    t=t+s[p];
                }
                //如果子串在最开头，子串后一位置都不能是字母
                if(k==0 && v[j]==t && (s[p]<'a' || s[p]>'z'))
                    info.n++;
                //如果子串出现在这行的其他位置,
                //那么，子串的前一位置和后一位置不能是字母
                else if(v[j]==t && (s[p]<'a' || s[p]>'z') & (s[k-1]<'a'
                    || s[k-1]>'z'))
                    info.n++;
            }
        }
        //字符串本身不能变，保持原样
        info.s=ss;
        vv.push_back(info);
    }
    sort(vv.begin(),vv.end(),comp);

    cout<<"Excuse Set #"<<line<<endl;
    for(i=0;i<vv.size();i++)
    {
        if(i!=0 && vv[i].n<vv[i-1].n)
            break;
        else
        {
            cout<<vv[i].s<<endl;
        }
    }
    //一个案例后面输出一个空行
    cout<<endl;
    }
    return 0;
}
```

4.48.7 汉语翻译

1. 题目

借口，借口！

Ito 法官遇到了一个棘手的问题，被传唤出庭做证的人为避免承担法律责任而找出漏洞百出的理由。为了减少听愚蠢的借口的时间，法官 Ito 要你编写一个程序，在一列借口中查找蹩脚的借口。借口关键字的匹配不区分大小写。

2. 输入描述

输入数据包含多个测试案例。

第 1 行包含两个整数。第一个数（$1 \leq K \leq 20$）定义了要查找的关键词的数量。第二个数（$1 \leq E \leq 20$）定义了被查找的借口的数目。

第 2 行到 $K+1$ 行每行包含一个关键词。

第 $K+2$ 行到 $K+1+E$ 行每行包含一句借口。

所有的关键词只包含连续的小写字母，每个关键词的长度是 L（$1 \leq L \leq 20$）。

所有的借口能包含任意大写和小写字母、空格或任何后跟的标点符号。如 [SPMamp".,!?&]，不包括方括号，长度不超过 70 个字符。

借口包含至少 1 个非空字符。

3. 输出描述

对于输入中的每个数据集，要求你打印出借口列表中最坏的借口。

最坏的借口被定义为包含最多的关键词。

如果一个关键词在一个借口中多次出现，每次都被算是一个独立的事件。

关键词"出现"在一个借口中，一定是这个关键词的开头与其他单词分开了，且它的后面是一行的结尾或是一个非字母的字符或空格。

对于每个输入的数据集，要求你在字符串"Excuse Set #"后边打印出数据集的序号（参见输出样例）。接下去打印出一行或多行最坏的借口，每个借口在一行上，打印的形式与读入的形式一样。如果有多个最坏的借口，你可以按任何顺序打印它们。

输出完一个测试案例后，你应当输出一行空行。

4. 输入样例

```
5 3
dog
ate
homework
canary
died
My dog ate my homework.
Can you believe my dog died after eating my canary... AND MY HOMEWORK?
This excuse is so good that it contain 0 keywords.
6 5
superhighway
crazy
thermonuclear
bedroom
war
building
I am having a superhighway built in my bedroom.
I am actually crazy.
1234567890....,,,,0987654321?????!!!!!!
There was a thermonuclear war!
I ate my dog, my canary, and my homework ... note outdated keywords?
```

5. 输出样例

```
Excuse Set #1
```

```
Can you believe my dog died after eating my canary... AND MY HOMEWORK?

Excuse Set #2
I am having a superhighway built in my bedroom.
There was a thermonuclear war!
```

4.49 Lowest Bit

4.49.1 链接地址

http://www.realoj.com/网上第 133 题

4.49.2 时空限制

Time Limit: 1000 ms   Resident Memory Limit: 1024 KB   Output Limit: 1024 B

4.49.3 题目内容

Given an positive integer A ($1 \leq A \leq 100$), output the lowest bit of A.

For example, given $A = 26$, we can write A in binary form as 11010, so the lowest bit of A is 10, so the output should be 2.

Another example goes like this: given $A = 88$, we can write A in binary form as 1011000, so the lowest bit of A is 1000, so the output should be 8.

Input

Each line of input contains only an integer A ($1 \leq A \leq 100$). A line containing "0" indicates the end of input, and this line is not a part of the input data.

Output

For each A in the input, output a line containing only its lowest bit.

Sample Input

```
26
88
0
```

Sample Output

```
2
8
```

4.49.4 题目来源

Zhejiang University Local Contest 2005 (Author: SHI, Xiaohan)

4.49.5 解题思路

本题是十进制到二进制的转换问题。十进制的整数转换为二进制的基本方法就是不断去除 2 并取余。

本题比较巧妙，不必把它的二进制先计算出来，只计算一部分即可。用十进制数不断去除 2，如果余数是 1 就终止，记录下来除 2 的次数 n，那么，2^n 就是我们需要输出的数据。因为其他位都是 0。

4.49.6 参考答案

```cpp
#include <fstream>
#include <iostream>
#include <cmath>
using namespace std;
int main(int argc, char * argv[])
{
    ifstream cin("aaa.txt");
    int n,num;
    while(cin>>n)
    {
        num=0;
        if(n==0)break;
        while(n%2!=1)
        {
            num++;
            n=n/2;
        }
        cout<<pow(2.0,num*1.0)<<endl;
    }
    return 0;
}
```

4.49.7 汉语翻译

1. 题目

最 低 位

给定一个整数 A（$1 \leqslant A \leqslant 100$），输出 A 的最低位。

比如，给定 $A=26$，我们可以写出 A 的二进制形式 11010，所以，A 的最低位就是 10，所以，你应当输出 2。

另一个例子是这样的：给定 $A=88$，我们写出 A 的二进制形式 1011000，所以，A 的二进制形式是 1000，所以，应当输出 8。

2. 输入描述

输入数据的每行包含一个整数 A（$1 \leq A \leq 100$）。$A=0$ 时，表示输入的结束，这行不要处理。

3. 输出描述

对于每个输入的 A，输出它的最低位在单独一行上。

4. 输入样例

```
26
88
0
```

5. 输出样例

```
2
8
```

4.50 Longest Ordered Subsequence

4.50.1 链接地址

http://www.realoj.com/ 网上第 134 题

4.50.2 时空限制

Time Limit: 1000 ms　Resident Memory Limit: 1024 KB　Output Limit: 1024 B

4.50.3 题目内容

A numeric sequence of a_i is ordered if $a_1 < a_2 < \cdots < a_N$. Let the subsequence of the given numeric sequence ($a_1, a_2, \cdots, a_N$) be any sequence ($a_{i1}, a_{i2}, \cdots, a_{iK}$), where $1 \leq i1 < i2 < \cdots < iK \leq N$. For example, the sequence (1, 7, 3, 5, 9, 4, 8) has ordered subsequences, e.g., (1, 7), (3, 4, 8) and many others. All longest ordered subsequences of this sequence are of length 4, e.g., (1, 3, 5, 8).

Your program, when given the numeric sequence, must find the length of its longest ordered subsequence.

Input

The first line of input contains the length of sequence N ($1 \leq N \leq 1000$). The second line contains the elements of sequence—N integers in the range from 0 to 10 000 each, separated by spaces.

Output

Output must contain a single integer—the length of the longest ordered subsequence of the given sequence.

This problem contains multiple test cases!

The first line of a multiple input is an integer N, then a blank line followed by N input blocks. Each input block is in the format indicated in the problem description. There is a blank line between input blocks.

The output format consists of N output blocks. There is a blank line between output blocks.

Sample Input

```
1

7
1 7 3 5 9 4 8
```

Sample Output

```
4
```

4.50.4 题目来源

Northeastern Europe 2002, Far-Eastern Subregion

4.50.5 解题思路

本题是一道经典的动态规划题目。动态规划（dynamic programming）是运筹学的一个重要分支，它是解决多阶段决策问题的一种有效的数量化方法。动态规划是由美国学者贝尔曼（R. Bellman）等人所创立的。1951年贝尔曼首先提出了动态规划中解决多阶段决策问题的最优化原理，并给出了许多实际问题的解法。1957年贝尔曼出版了《动态规划》一书，标志着运筹学这一重要分支的诞生。

动态规划可以优雅而高效地解决许多用贪心算法或分治算法难以解决的求最优解的问题。而求最优解的问题往往看似简单，但程序实现起来很难，往往容易出现重复计算，产生超时错误。

用动态规划解题，首先要把原问题分解为若干个子问题，这一点与递归方法类似。动态规划与递归的区别在于：单纯的递归往往会导致子问题被重复计算，而运用动态规划方法，子问题的解一旦被求出就会被保存，所以，每个子问题只需求解一次。

本题是要求最长上升子序列，如何把这个问题分解为子问题呢？经过分析，发现"求以 $a_k(k=1,2,3,\cdots,N)$ 为终点的最长上升子序列的长度"是个比较好的子问题（这里把一个上升子序列中最右边的那个数称为该子序列的终点）。a_k 是测试案例中的每一个元素。只要把这 N 个子问题解决了，那么这 N 个子问题的解中，最大的那个就是整个问题的解。

上述子问题只与一个变量相关，即数字的位置。因此序列中数字的位置 k 就是状态，而状态 k 对应的值，就是以 a_k 作为终点的最长上升子序列的长度。这个问题的状态一共有

N 个。状态定义出来后,状态转移方程就比较好写出来了。假定 MaxLen(k) 表示以 a_k 作为终点的最长上升子序列的长度,那么状态转移方程为:

$$\begin{cases} \text{MaxLen}(1) = 1 \\ \text{MaxLen}(k) = \text{Max}\{\text{MaxLen}(i) : 1 \leqslant i < k \text{ 且 } a_i < a_k \text{ 且 } k \neq 1\} + 1 \end{cases}$$

这个状态转移方程的意思是,MaxLen(k) 的值,就是在 a_k 左边,终点数值小于 a_k,且长度最大的那个上升子序列的长度再加 1。因为 a_k 左边任何终点数值小于 a_k 的子序列,加上 a_k 后就能形成一个更长的上升子序列。

有了上述的状态转移方程就可不必编写递归函数,因为从 MaxLen(1) 可以推出 MaxLen(2),有了 MaxLen(1) 和 MaxLen(2) 就能推出 MaxLen(3),直到推出 MaxLen(N),而这当中,子问题在求解后,其解都作了保存,所以,子问题是不会被重复计算的,时间就省在这里了。

在使用泛型编程时,定义结构体 Info 来保存每个子问题的计算结果:

```
struct Info{
        int index;          //该数据的序号,1, 2, 3, …, n
        int a;              //该数据
        int length;         //以该数据为终点的最长子系列长度
};
```

而每个子问题的结果都作为 vector 向量的一个元素,这里特别注意的是,如果向量的元素类型是结构体,那么,要先使用 push_back() 方法先增加一个结构体空间,再对该元素的结构体变量的各分量赋值。

再有一点需注意的是,输出块之间有一空行,而不是每一块输出块后面跟一空行。

4.50.6 参考答案

```
#include <vector>
#include <iostream>
#include <fstream>
using namespace std;
//定义结构体Info,用来保存一个终点数据的信息
struct Info{
    int index;//该数据的序号,1, 2, 3, …, n
    int a;//该数据
    int length;//以该数据为终点的最长子系列长度
};
vector<Info> v;//定义向量v来保存所有的终点数据信息
int main(int argc, char * argv[])
{
    ifstream cin("aaa.txt");
    Info info;
    //测试案例个数为c
    int c;
    //每个测试案例的个数
    int t;
```

```cpp
//单个元素
int e;
//元素计数器
int n;
//子序列的最大长度
int m;
int i,j,k,p;
cin>>c;
for(i=0;i<c;i++)//共有c组测试案例
{
    //清空向量v
    v.clear();
    //初始化元素计数器
    n=0;
    //读入每个测度案例的个数
    cin>>t;
    //开始计算一个测试案例
    for(j=0;j<t;j++)
    {
        cin>>e;//读入一个元素
        n++;//元素数量加1
        //一定要先增加一个结构体元素，再对结构体的各变量赋值
        v.push_back(info);
        v[n-1].index=n;
        v[n-1].a=e;
        if(n==1)//是第一个元素
        {
            v[n-1].length=1;
            continue;
        }
        //找出位于本元素之前,比本元素小但子序列最长的那个长度值
        m=0;
        for(k=0;k<n;k++)
        {
            if(v[k].a<e)
            {
                if(v[k].length>m)m=v[k].length;
            }
        }
        v[n-1].length=m+1;
    }
    //找出本组测试案例中最长子序列的长度值
    m=0;
    for(p=0;p<n;p++)
    {
        if(v[p].length>m)m=v[p].length;
    }
```

```
        //输出结果,注意,输出块之间有一空行
        if(i!=0)cout<<endl;//不是第一行,先输出一行空行
        cout<<m<<endl;
    }
    return 0;
}
```

4.50.7 汉语翻译

1. 题目

最长上升子序列

一个数的序列 a_i,当 $a_1 < a_2 < \cdots < a_N$ 时,称这个序列是上升的。对于给定的一个序列 $(a_1, a_2, \cdots, a_N)$,可以得到一些上升子序列 $(a_{i1}, a_{i2}, \cdots, a_{iK})$,这里 $1 \leq i1 < i2 < \cdots < iK \leq N$。例如,对于序列(1,7,3,5,9,4,8),它包含一些上升子序列,如(1,7),(3,4,8)等。这些子序列中最长的长度是4,比如子序列(1,3,5,8)。

你的任务是,对于给定的数字序列,求出最长上升子序列的长度。

2. 输入描述

输入数据的第一行是序列的长度 N($1 \leq N \leq 1\,000$)。第二行包含序列的元素(N 个整数,每个元素的大小是从 0 到 10 000),中间用空格分开。

3. 输出描述

输出必须包含一个整数(这个整数是给定序列的最长上升子序列的长度)。
本问题包含多个测试案例!
输入数据的第一行是一个整数 N,然后是一行空行,后边是 N 个输入块。每个输入块的格式在问题描述中已说明了。每个输入块之间有一空行。
输出格式包含 N 个输出块。输出块之间要有一行空行。

4. 输入样例

```
1

7
1 7 3 5 9 4 8
```

5. 输出样例

```
4
```

附录1　用 VC++编写控制台程序的方法

在 ACM 竞赛中，可以使用 Microsoft Visual C++ 6.0（SP6）来编写程序，主要的好处是会自动弹出 STL 对象的方法。下面讲讲 VC++ 6.0 的控制台程序的编写方法。

例题

编制一个 C++程序，输入 a 和 b 两个整数，输出这两个整数的和。

操作

（1）运行 VC++ 6.0，单击 File | New 菜单项，再在弹出的 New 对话框的 Projects 选项卡中单击 Win32 Console Application（Win32 控制台应用程序），如图 1 所示。

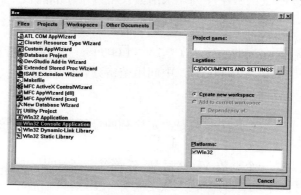

图 1　选定 Win32 控制台应用程序

（2）单击 New 对话框右上方 Location（位置）下边的 ... 按钮，在弹出的 Choose Directory 对话框中单击文件夹列表中的"桌面"后再单击 OK 按钮，就把工程的位置设定在桌面上，即 C:\DOCUMENTS AND SETTINGS\ADMINISTRATOR\桌面\，如图 2 所示。

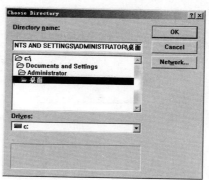

图 2　选择桌面位置

（3）再在 New 对话框的右上方的 Project name 文本框中输入工程名称"1_2"，如图3所示。

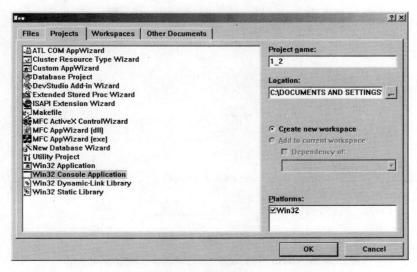

图3　输入工程名称"1_2"

（4）单击 OK 按钮，在弹出的向导对话框中选中 A simple application.单选按钮（一个简单的应用程序），如图4所示。

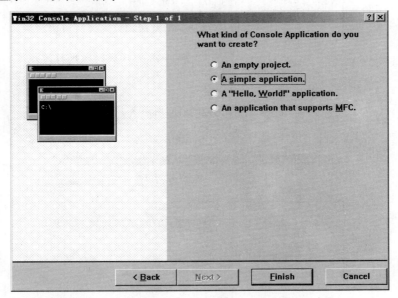

图4　选择 A simple application

（5）单击 Finish（完成）按钮后，弹出 New Project Information 对话框，如图5所示。

（6）单击 OK 按钮，这样，一个简单的控制台应用程序 1_2.cpp 就创建了，如图6所示。

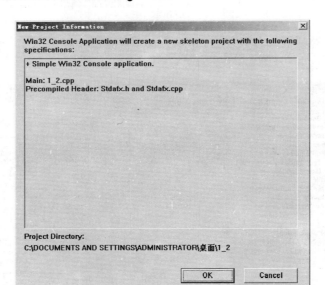

图 5 创建文件列表

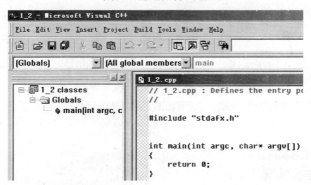

图 6 创建了简单的控制台应用程序 1_2.cpp

（7）现在，在 1_2.cpp 中直接编写代码，如图 7 所示。

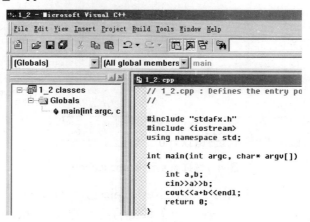

图 7 编写代码

小提示

在 VC++ 6.0 中，C++类都在 std 命名空间中，所以，如果是编写 C++程序，那么，都需要使用"using namespace std;"语句来声明程序中的 C++类是在 std 命名空间中，否则，程序会出现编译错误。

标准输入流对象 cin 和标准输出流对象 cout 在头文件 iostream 中定义了"extern _CRTIMP istream cin;"和"extern_CRTIMP ostream cout;"，所以需要头文件包含声明"#include <iostream>"。cin 默认的对象是键盘设备，cout 默认的对象是屏幕设备。

另外，包含 C++文件的方法都是采用"#include <iostream>"的形式。在 VC++ 6.0 中，C++类文件名都不带".h"，而带".h"的头文件名称都是 C 语言的。C/C++的头文件位置在 C:\Program Files\Microsoft Visual Studio\VC98\Include 文件夹中，如图 8 所示。

图 8　C++类头文件名称都不包含".h"

（8）单击工具栏中的 ! 按钮，或者按 Ctrl + F5 组合键来编译运行本程序，这时，会询问是否生成目标文件和可执行文件，如图 9 所示。

（9）单击"是"按钮，就开始编译 1_2.cpp，编辑界面最底部的状态显示为"1_2.exe - 0 error(s), 0 warning(s)"，即出现 0 个错误，0 个警告，说明程序不存在语法错误。然后，自动弹出 1_2.exe 文件运行控制台，我们在输入"1"后按回车键，再输入"2"后按回车键，立即会输出 1+2 的结果"3"，如图 10 所示。

（10）按任意键，控制台窗口会自动关闭，回到编辑状态。

图 9 是否生成目标文件和可执行文件

图 10 程序运行控制台

（11）退出 VC++ 6.0。双击桌面上的 1_2 文件夹，可以看到工程的全部文件，如图 11 所示。

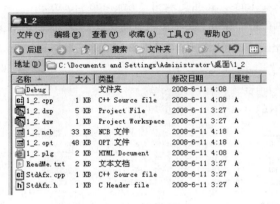

图 11 1_2 工程文件

（12）编译后的可执行程序是 1_2\Debug\1_2.exe，双击它，就可运行程序；而 1_2\1_2.dsw 是工程文件，双击它，就把 1_2 工程打开了。

（13）本操作到此结束。

小提示

Visual Studio .NET 2008 的 C++ 头文件在 C:\Program Files\Microsoft Visual Studio 9.0\VC\Include 文件夹中。

如果使用 Visual Studio .NET 2008 的 VC++ 来编写 Win32 控制台程序，那么，所有 C++STL 头文件包含语句和 "using namespace std;" 语句需要写在 stdafx.h 文件里，否则，会出现编译错误。

附录 2 本书试题第三方 ACM 网站链接

附表 1 第 3 章试题在浙江工业大学 ACM 网站的链接地址

题 号	链 接 地 址
3.1　读入一个参数	http://acm.zjut.edu.cn/网上第 1167 题
3.2　读入两个参数	http://acm.zjut.edu.cn/网上第 1166 题
3.3　1!到 n!的和	http://acm.zjut.edu.cn/网上第 1174 题
3.4　等 比 数 列	http://acm.zjut.edu.cn/网上第 1175 题
3.5　菲波那契数	http://acm.zjut.edu.cn/网上第 1176 题
3.6　最大公约数	http://acm.zjut.edu.cn/网上第 1177 题
3.7　最小公倍数	http://acm.zjut.edu.cn/网上第 1178 题
3.8　平 均 数	http://acm.zjut.edu.cn/网上第 1179 题
3.9　对称三位数素数	http://acm.zjut.edu.cn/网上第 1181 题
3.10　十进制转换为二进制	http://acm.zjut.edu.cn/网上第 1185 题
3.11　列 出 完 数	http://acm.zjut.edu.cn/网上第 1190 题
3.12　12! 配 对	http://acm.zjut.edu.cn/网上第 1191 题
3.13　五位以内的对称素数	http://acm.zjut.edu.cn/网上第 1187 题
3.14　01 串 排 序	http://acm.zjut.edu.cn/网上第 1204 题
3.15　排列对称串	http://acm.zjut.edu.cn/网上第 1208 题
3.16　按绩点排名	http://acm.zjut.edu.cn/网上第 1205 题
3.17　按 1 的个数排序	http://acm.zjut.edu.cn/网上第 1044 题

附表 2 第 4 章试题在浙江大学 ACM 网站的链接地址

题 号	链 接 地 址
4.1　Quicksum	http://acm.zju.edu.cn/网上第 2812 题
4.2　IBM Minus One	http://acm.zju.edu.cn/网上第 1240 题
4.3　Binary Numbers	http://acm.zju.edu.cn/网上第 1383 题
4.4　Encoding	http://acm.zju.edu.cn/网上第 2478 题
4.5　Look and Say	http://acm.zju.edu.cn/网上第 2886 题
4.6　Abbreviation	http://acm.zju.edu.cn/网上第 2947 题

续表

题　　号	链接地址
4.7　The Seven Percent Solution	http://acm.zju.edu.cn/网上第2932题
4.8　Digital Roots	http://acm.zju.edu.cn/网上第1115题
4.9　Box of Bricks	http://acm.zju.edu.cn/网上第1251题
4.10　Geometry Made Simple	http://acm.zju.edu.cn/网上第1241题
4.11　Reverse Text	http://acm.zju.edu.cn/网上第1295题
4.12　Word Reversal	http://acm.zju.edu.cn/网上第1151题
4.13　A Simple Question of Chemistry	http://acm.zju.edu.cn/网上第1763题
4.14　Adding Reversed Numbers	http://acm.zju.edu.cn/网上第2001题
4.15　Image Transformation	http://acm.zju.edu.cn/网上第2857题
4.16　Beautiful Meadow	http://acm.zju.edu.cn/网上第2850题
4.17　DNA Sorting	http://acm.zju.edu.cn/网上第1188题
4.18　Daffodil Number	http://acm.zju.edu.cn/网上第2736题
4.19　Error Correction	http://acm.zju.edu.cn/网上第1949题
4.20　Martian Addition	http://acm.zju.edu.cn/网上第1205题
4.21　FatMouse' Trade	http://acm.zju.edu.cn/网上第2109题
4.22　List the Books	http://acm.zju.edu.cn/网上第2727题
4.23　Head-to-Head Match	http://acm.zju.edu.cn/网上第2722题
4.24　Windows Message Queue	http://acm.zju.edu.cn/网上第2724题
4.25　Language of FatMouse	http://acm.zju.fu.cn/网上第1109题
4.26　Palindromes	http://acm.zju.edu.cn/网上第2744题
4.27　Root of the Problem	http://acm.zju.edu.cn/网上第2818题
4.28　Magic Square	http://acm.zju.edu.cn/网上第2835题
4.29　Semi-Prime	http://acm.zju.edu.cn/网上第2723题
4.30　Beautiful Number	http://acm.zju.edu.cn/网上第2829题
4.31　Phone List	http://acm.zju.edu.cn/网上第2876题
4.32　Calendar	http://acm.zju.edu.cn/网上第2420题
4.33　No Brainer	http://acm.zju.edu.cn/网上第2201题
4.34　Quick Change	http://acm.zju.edu.cn/网上第2772题
4.35　Total Amount	http://acm.zju.edu.cn/网上第2476题
4.36　Electrical Outlets	http://acm.zju.edu.cn/网上第2807题
4.37　Speed Limit	http://acm.zju.edu.cn/网上第2176题

续表

题 号		链接地址
4.38	Beat the Spread!	http://acm.zju.edu.cn/网上第 2388 题
4.39	Champion of the Swordsmanship	http://acm.zju.edu.cn/网上第 2830 题
4.40	Doubles	http://acm.zju.edu.cn/网上第 1760 题
4.41	File Searching	http://acm.zju.edu.cn/网上第 2840 题
4.42	Old Bill	http://acm.zju.edu.cn/网上第 2679 题
4.43	Divisor Summation	http://acm.zju.edu.cn/网上第 2095 题
4.44	Easier Done Than Said?	http://acm.zju.edu.cn/网上第 1698 题
4.45	Let the Balloon Rise	http://acm.zju.edu.cn/网上第 2104 题
4.46	The Hardest Problem Ever	http://acm.zju.edu.cn/网上第 1392 题
4.47	Fibonacci Again	http://acm.zju.edu.cn/网上第 2060 题
4.48	Excuses, Excuses!	http://acm.zju.edu.cn/网上第 1315 题
4.49	Lowest Bit	http://acm.zju.edu.cn/网上第 2417 题
4.50	Longest Ordered Subsequence	http://acm.zju.edu.cn/ 网上第 2136 题

参 考 文 献

[1] 钱能. C++程序设计教程（第二版）实验指导[M]. 北京：清华大学出版社，2007.
[2] 叶至军. C++STL开发技术导引[M]. 北京：人民邮电出版社，2007.
[3] 李文新，等. 程序设计导引及在线实践[M]. 北京：清华大学出版社，2007.